I0150990

AIR OPERATIONS 1958
LEBANON AND TAIWAN

By Jacob Van Staaveren,
Robert D. Little and Wilhelmina Burch

DEFENSE
LION
PUBLICATIONS

Acknowledgements

Air Operations in the Taiwan Crisis of 1958 and Air Operations in the Lebanon Crisis of 1958 are parts of a series of studies on air operations in international incidents, prepared by the USAF Historical Division Liaison Office at the request of the Directorate of Plans, Headquarters USAF. These historical narratives are based on primary source materials available in 1960 -- messages and correspondence - and on histories from many levels of the Air Force, units, commands, and the Air Staff. Originally prepared in a very few copies, these studies have been in great demand by the Air Staff, major commands, and Department of Defense agencies and are now being published for wider dissemination. This book is a compilation of those reports with substantial additional material that updates and complements the original material. However, the integrity of the original text has been scrupulously preserved.

Copyright Notice

Copyright © 2012 Lion Publications Inc, 22 Commerce Road, Newtown, Connecticut 06470. ISBN 978-0-9859730-3-2 No part of this compilation may be reproduced or transmitted in any form or by any means, electronic or mechanical photocopying, recording or by any information and retrieval system without permission in writing from the publisher

ORIGINAL FOREWORD

The defense of Taiwan has been and will continue to be a source of major concern to the United States and especially to the U.S. armed services. It is likely that there may be more military crises involving Taiwan and its related islands. The crisis in 1958 provided a test of American military planning concepts that should prove of value for future planning. Occurring in the Pacific almost simultaneously with the Lebanon crisis of July-August 1958, it created certain planning, operational, and logistic problems that had not been anticipated.

[The study on the Lebanon] written by Dr. Robert D. Little and Mrs. Wilhelmina Burch is of special significance in connection with planning and. preparations for local wars and incidents. In this operation. the American forces experienced no combat action, but the possibility of action was constant and the deployment of forced to the objective area was of first importance. To achieve a meaningful historical context for the military operations, the political and diplomatic background has been presented in some detail.

JOSEPH W ANGELL, JR.
Chief, USAF Historical Division Liaison Office

****

2012 FORWARD
By Defense Lion Publications

More than fifty years have passed since the twin crises of Lebanon and Taiwan dominated the news. These two events have faded into history and are often regarded as being, at best, footnotes in the histories of the Cold War. The ships that took part in them have long since gone to the scrapyard, the aircraft and the veterans who flew them have retired. The world in which they took place seems to have passed away a long, long time ago. Yet, when reading these monographs, it become apparent that these two crises were profoundly important in terms of the effects they were to have on American strategy throughout the rest of the century. Those effects are still in play today and even now they are affecting how the United States perceives the world and reacts to developments within it. In strategic terms, a very short line links the twin crises of 1958 to the Vietnam War and the connection to Operation Desert Storm and Operation Iraqi Freedom is not that much longer. It is no coincidence that those three military operations took place in the same area as the 1958 crises.

In general terms, the two crises of 1958 were strikingly similar. In both cases, a long-simmering stew of regional tensions and local political ambitions boiled over into a violent confrontation. In both cases, the United States believed that its regional interests required it to intervene in order to stabilize the situation and it did so by moving in relatively small but capable military forces that established a trip-wire. This served notice to the other participants that any expansion of the conflict beyond its existing limits would result in a major military confrontation with the United States. In both cases, the deployment was (in the short term at least) successful and the situations both stabilized. U.S. strategic mobility was decisive in both conflicts; it enabled aircraft to be flown to the scene of operations and supplied once there. Troops could be lifted in by air or landed from the sea. However, there were limits to what could be achieved by such strategic mobility. Tactical aircraft could fly to the scene of operations but they would not be combat-ready when they arrived. Troops could be flown to a trouble spot but they would need time to reorganize and prepare for operations once they got there. The hard truth was that the U.S. Navy's aircraft carriers and Marines arrived as organized and supplied forces that were a much better way of responding to such crises.

In 1958, on paper, the crises in Lebanon and Taiwan were handled 'on the margin' without any significant effects on the primary missions facing the U.S. force structure. That was on paper. The reality was somewhat different in that both deployments caused disruptions that extended beyond the forces that were actually used. The force levels really deployed to the areas actually represented a much greater proportion of the total force level actually available than it appeared. Lebanon and Taiwan were sequential crises; some of the assets used in the latter had also taken part in the former. If the crises had been simultaneous, the U.S. would have had greater problems in dealing with them. From this demonstration grew the concept that the U.S. had to have a realistic, not theoretical, ability to handle two regional crises simultaneously. Today, in 2012, that capability carefully maintained for half a century, is on the verge of being sacrificed.

Prior to the 1958 crises, U.S. strategy had been based on the assumption that the use of nuclear weapons would dominate future wars. The deployments to Lebanon and Taiwan suggested that this assumption was flawed. In both cases, nuclear-capable assets were scheduled for deployment but it quickly became apparent that they were inappropriate to the situation. This caused momentary embarrassment in the Lebanon where a nuclear-capable battery of Honest John missiles had to be hurriedly returned but in Taiwan, the optimization of the force structure for a nuclear war was a serious problem. The ultimate impact of these considerations led to the abandonment of 'massive retaliation' as a U.S. strategy and the institution of a more nuanced 'flexible response'. This led to the revitalization of American conventional forces and the establishment of a much more capable U.S. non-nuclear defense structure. It was the U.S. Army that was formed from the lessons of Lebanon and Taiwan that went to Vietnam.

CONTENTS

MAPS AND CHARTS

PART ONE – AIR OPERATIONS IN LEBANON

THE EASTERN MEDITERRANEAN

TURKEY

INCIRLIK AIR BASE ★ ● ADANA

ISKENDERUN

NICOSIA ●

CYPRUS

SYRIA

TRIPOLI ●

Mediterranean Sea

BEIRUT ●

INTERNATIONAL AIRPORT ★

LEBANON

● DAMASCUS

AMMAN ●

JERUSALEM ●

PORT SAID

ISRAEL

JORDAN

CAIRO ●

E G Y P T

0   77   1 IN = 77 STATUTE MILES

The Lebanon Area In 1958

1

I .. THE CRISIS IN LEBANON

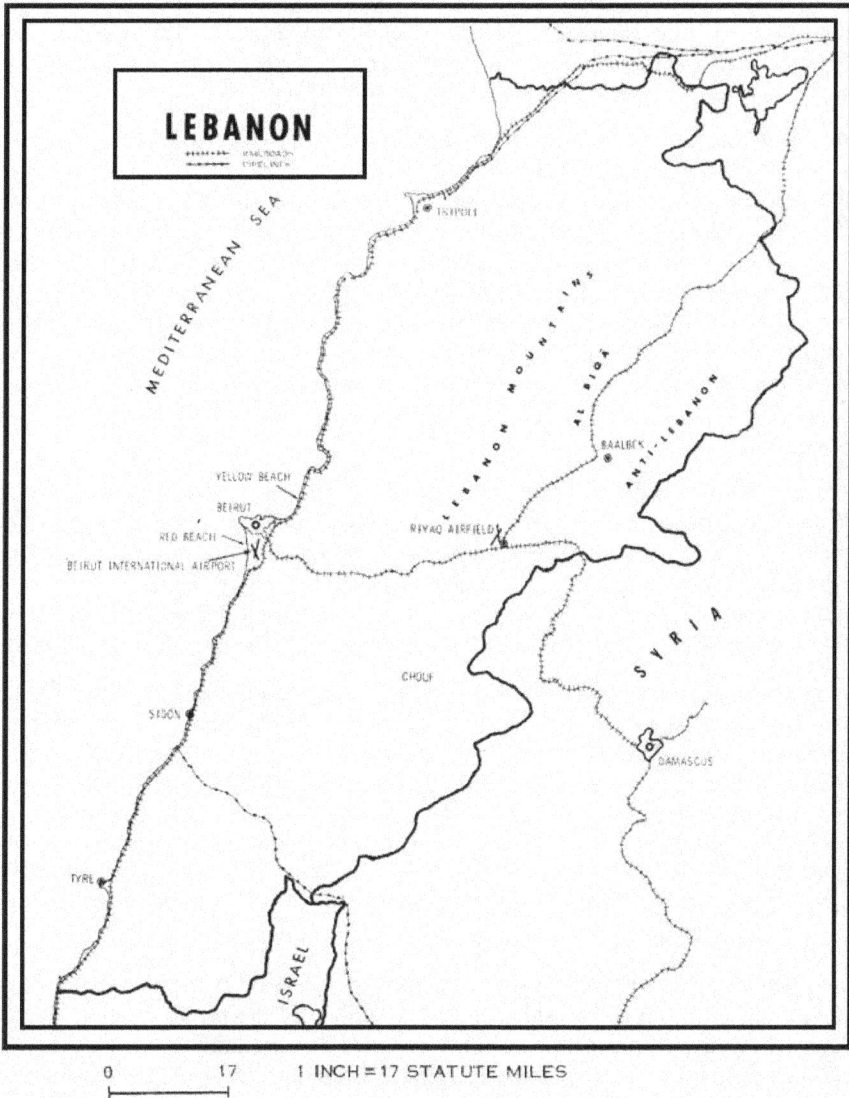

LEBANON

0   17   1 INCH = 17 STATUTE MILES

Lebanon is a small country of about 4,000 square miles. Except for a short common boundary with Israel to the south and a sea frontage on the Mediterranean, it is surrounded. by its much larger neighbor, Syria which in February 1958 joined with Egypt to form the United Arab Republic.

The population of Lebanon, like that of the other Levant countries, is highly mixed in nature, but it is unique among the countries of the Middle East in that no one ethnic group constitutes a decided majority. Christians and Mohammedans are roughly about equal in number, with both claiming a preponderance which the lack of a formal census since 1942 makes impossible to establish. Because of much heavier emigration by the Christians and. a higher rate of natural increase by the Moslems, it appears inevitable that the latter will soon constitute a majority, if they have not already done so. Further complicating the picture are more than 100,000 Palestinian refugees, of whom 80 percent are Moslem. These entered the country during the Israel-Arab war of 1948-49 in which Lebanon participated.

The most numerous of the Christian sects are the Maronites, who in 1958 numbered about 422,000 of the approximate 1,500,000 population. Like several other Uniate groups, the Maronites recognize the primacy of the Pope but retain privileges and procedures of their own. Other important Christian sects are the Greek Orthodox (198,000), the Greek Catholics (90,000), the Armenian Orthodox (58,000), and the Armenian Catholics (14,000). The most numerous Moslem sects are the Sunnis, or Sunnites, also considered the most orthodox, who number about 274,000; the Shiahs, or Shiites (225,000); and the Druzes (88,000), the last-named regarded as pronounced. heretics. In many cases these Christian and Moslem sects are tightly bound social groupings, sometimes located in long-established geographical enclaves. [1]

Lebanon, with an estimated population of about 400 per square mile, is a crowded country and becoming increasingly so. Arabic is the principal and official language; French is an important secondary language. Educationally, the country is far ahead of the other official Eastern Arab countries. Two thirds of the school-age population are actually attending, and several colleges exist in Beirut, including the well-known American University founded by missionaries in 1866.

Long a part of the Turkish Empire, Lebanon was declared a League of Nations mandate after World War I and placed under the French. It was granted a constitution and became officially a republic in 1925 but actually remained under close French control, receiving full independence only in 1944. The French period served to accelerate the Westernization and modernization of the country to a degree unsurpassed in the Arab Middle East. Nevertheless, the country remained politically immature, and true political parties existed only in embryo form. After the 1953 election no party controlled more than four votes, and the great majority of the deputies in the single-house legislature were independents.

The attempt to preserve a religious balance in the country was reflected in the political custom that the president be a Maronite and the prime minister a Sunni Moslem. By custom also, the cabinet included representatives of various other religious faiths. In addition to the religious differences, the country was divided, by the advocacy of Western or Arab orientation in government, not always following strictly religious lines.[2]

The earliest Lebanese parties, appearing in the 1930's, were the Constitutional Bloc, led by Bishara al-Khuri, and the National Bloc, led by Emile Edde. The principal difference between the two perhaps lay in the fact that Khuri, although a Maronite, proposed a policy of maximum cooperation with Arab countries and a struggle against Western "imperialism" while Edde, who had served as president under the French mandate, opposed identification with the Arab countries and advocated a Western orientation of policy. By 1952, Khuri, who was serving as president, found himself in a position in which he had lost much support because of revelations of inefficiency and malfeasance in his government and was virtually forced to resign by popular demand.. He was succeeded in September 1952 by Camille Chamoun, a Maronite leader with Sunni Moslem support, who later secured, a large majority in the legislature elected in 1953.[3]

During the six years of his term Chamoun faced increasing opposition within the country, particularly from Moslem elements of the population, largely as a result of the emergence of Nasser and the incorporation of Syria within the United Arab Republic. The pan-Arabic Moslem elements objected. to the Chamoun government's Western orientation of policy and demanded a larger share of political control. It is obvious that Chamoun, while maintaining a large majority of support in the legislature, developed. serious personal differences with various influential leaders, both Moslem and Christian, During the same period there was a rapid growth in organized political movements of various sorts. In May 1957, Chamoun again won an overwhelming majority of support in the elections for the assembly.[4]

Camille Chamoun
Source: Lebanese
Government

The May Rebellion

Agitation against Chamoun culminated on 9 May 1958 in the outbreak of an armed rebellion. The rebels quickly secured control of most of the outlying sections of the country as well as considerable parts of the cities of Beirut and Tripoli. They were plentifully supplied with arms, apparently from Syrian sources, and Syrian volunteers also entered the country to join the rebel forces. The rebel leaders were well supplied with funds, probably received from Nasser, enabling them to recruit additional forces from both local and foreign sources by paying a daily wage. Despite these advantages, the rebels failed to make important headway after the first week of the uprising, and a condition of stalemate persisted thereafter. [5]

There were several reasons for this. Although strongly pan-Arabic in inclination, the rebel leaders had some dissimilar aims, their common denominator being opposition to Chamoun. The latter also had solid sources of support, as will be explained, including considerable armed groups or partisans. Furthermore, the national army, though small, was loyal and well disciplined and proved more than a match for the rebels in open conflict. However, General Fouad Chehab, its commander, contended that its various religious composition made it too brittle an instrument to employ in an all-out campaign against the rebels and proposed to let the rebellion, once contained, die out of itself. Undoubtedly an important factor in the minds of both sides was that Chamoun's term of office would end in September 1958. Presumably a compromise successor could compose differences and end the stalemate.[6]

Saeb Salaam
Source: Lebanese Government

The pro-government and anti-government groupings were quite complex, in some cases cutting across religious lines, Supporting the revolt were such political-organizations as the Progressive Socialists, principally Druze in composition; Njjadah, the militant pan-Arabic youth movement; the Muslim Brotherhood, extreme Islamic; and the League of National Action, milder pan-Arabic. More important were the personal followings of the principal leaders. Saeb Salaam, political boss of the Basta, the Moslem section of Beirut, was perhaps the most extreme. Kamal Jumblatt, a feudal aristocrat who had founded the Progressive Socialist Party and given away part of his family lands, held sway among the Druzes of the Chouf, an area some miles south east of the capital. Rashid Kerani, leader of the Tripoli opposition, was considered relatively moderate. A 31-year-old lawyer educated in Cairo, he was a Sunni Moslem deputy who had served for a time during 1955-56 as prime minister under Chamoun.[7]

Probably the strongest organized. political group supporting Chamoun was the Phalange (Les Phalanges Libanaises), militant nationalists who were predominantly Maronites though officially disclaiming confessional leanings. Led by Pierre Gemayel, they dominated the Ashrafiye, the Christian quarter of Beirut, and fielded an armed body of partisans organized along military lines. Chamoun was also supported. by the PPS (Parti Populaire Syrien), an armed and organized youth group that had been outlawed in 1941 because of its belligerency. Standing for complete separation of church and state, it placed Syrian nationalism (including Lebanon as a part) ahead of the pan-Arabic movement. It had thus opposed absorption of Syria within the UAR and now opposed the revolt led by Nasser's followers in Lebanon. Chamoun himself, despite his large popular and legislative following, led no organized party, though he did organize the National Liberal Union Party before leaving the presidency in September 1958. Further confusing the picture was the support of Chamoun by the Sunni Moslem prime minister, Sami Solh, and the bitter opposition to Chamoun by the Maronite Patriarch Meouchi, who advocated appeasement of Nasser.[8]

Between the extremes were various figures who advocated compromise and who were sometimes loosely grouped together as the Third Force (also ironically termed. the Third. Farce by its critics). These included Henri Pharaon, a wealthy Greek Orthodox landowner and financier, and. Raymond Edde, son of the former president under the French mandate. Leader of the largest party in the national assembly, the National Bloc, which controlled 4 of the 66 seats, Edde became principal leader of the Third Force movement. He was later to become a key figure in working out a compromise cabinet. Even more important than these was General Chehab, a Maronite, who had, served briefly as provisional president in 1952 and now as commander of the army advocated a policy of compromise that would minimize bloodshed. Generally acceptable to all sides as a compromise president, he seemed reluctant to allow himself to be named.[9]

Raymond Edde
Source: Lebanese Government

During the entire period of late May, June, and early July the situation remained essentially static, despite sporadic violence. Bombing, sniping, and occasional armed dashes took place, but there were no extensive and. prolonged. aggressive movements by either side. Street barricades defended the large sections of Beirut and Tripoli that were in opposition hands, and the army made no sustained efforts to take them. At the same time, the rebels made no important headway towards welding their forces throughout the country into a cohesive whole. The Lebanese Air Force, composed of 15 jet planes (mostly British Vampires) and 18 pilots, helped to prevent such moves on the few occasions when rebel forces, particularly Jumblatt's Druzes, ventured into open country.

General Fuad Chehab
Source: Lebanese Government

As it became obvious that fighting might continue indefinitely and the rebels in Tripoli suffered heavy casualties in attempts to take over complete control of the city, the enthusiasm of Syrian volunteers for joining the rebels apparently waned. 0n the other hand, some desertions occurred from both the army and gendarmerie.

Chamoun's support in the National Assembly and among most of the Christian elements of the population remained firm, but his Moslem support undoubtedly weakened, as the result of pressure of all kinds exerted by the opposition, including the assassination of several minor leaders and threats to major ones. Nevertheless, Chamoun continued to receive some Moslem support, since this was frequently tied to personalities, and Moslem leaders like Sarni Solh remained in the cabinet.[10]

The Lebanese government decided to place its problem before the Arab League, the regional organization of Arab states. At a meeting in Bengasi, Libya, beginning 31 May, the League considered the Lebanese complaint that the United Arab Republic had instigated and was supporting the revolt. The League took no action, since the United Arab Republic refused to accept even a mildly worded resolution of 5 June 1958, which the Lebanese government itself found completely inadequate. [11]

Meanwhile, on 22 May, Lebanon had requested a meeting of the Security Council of the United Nations to consider its charge that the United Arab Republic had instigated the rebellion and was supporting it with arms and "volunteers," but, later, Lebanon asked that the meeting be postponed to permit prior consideration by the Arab League. When it became clear that nothing effective could be expected from that source, the Security Council meeting was scheduled for 11 June. [12]

Although the UAR representative denied the Lebanese charges, the Security Council voted to adopt a Swedish resolution, which was supported by both the United States and United. Kingdom, that a UN observer group be dispatched to Lebanon to report on outside intervention. The vote was 10 to 0, the Soviet Union abstaining. The vanguard of the UN observer group consisting of five men, arrived in Lebanon the following day, 12 June, in the midst of an intensified phase of the fighting. By 18 June, about 50 UN observers were in Lebanon, with about 75 more expected soon. The following day Dag Hamarskjold, Secretary General of the United Nations, arrived in Lebanon for a two-day visit during which he conferred with President Chamoun and received preliminary reports from the observers. [13]

Dag Hammarskjold
Source: U.N. Archives

The arrival of the UN observers and the visit by Hamarskjold apparently had some effect in quieting the fighting. Another effect noted was the replacement on the street barricades in Beirut of UAR flags and pictures of Nasser by Lebanese flags and pictures of rebel leaders. The influence of the observers, who had risen to about 100 men by 29 June, was essentially psychological, since it was impossible in most cases to gain access to the rebel-held frontier areas, where arms and recruits were most likely to be moved in. Nevertheless, on 4 July, Secretary General Hammsarskjold ventured to state that the charge of "massive infiltration" was not warranted at that time. The following day the first report of the UN observer group, which was headed by former president Galo Plaza of Ecuador, stated in essence that no significant movement of personnel or supplies into the country had been observed and that the rebellion must be regarded as a civil conflict. Galo Plaza admitted, however, that his group had been able to gain access to only 18 of the 278 kilometers of frontier.[14]

Sporadic violence continued to erupt, but there were indications that the rebel factions were tiring of the struggle and inclined to seek a compromise that would involve no serious loss of face. Increasing attention was focused on the scheduled election of a new president by the parliament on 24 July. The economic repercussions of the rebellion increasingly pinched the business community, and the executive council of the Lebanese Industrial Association passed. a unanimous resolution on 9 July threatening to close all factories unless the election was held as scheduled. and a new president chosen. The Lebanese constitution forbade the reelection of a president but at the same time gave him very great powers. After the victory of his supporters in the election of May 1957, Chamoun was accused by the opposition of planning to amend the constitution to permit his reelection and even of planning to continue in office illegally. After the outbreak of violence in May 1958 Chamoun stated his intention to leave office at the expiration of his term on 23 September. There is little doubt, however, that some of his supporters wanted to postpone the presidential election until after the suppression of the rebellion. The opposition, on the other band, demanded that Chamoun step out as a first condition for negotiating a settlement. Actually, the rebel opposition appears to have felt that it was insufficiently represented in the national assembly, which elected the president, and hoped to influence it to select a man not aligned with Chamoun.[15]

The Decision to Send Military Aid

The uprising in Lebanon in May 1958 eventually had far-reaching consequences for the United States. In previous statements of policy in 1956 and 1957, the United. States had served notice that it was "prepared. to use armed forces" in response to the appeal of any victim of armed aggression in the Middle East. The American policy of supporting stable, friendly, progressive governments in the area was directly challenged by the rebellion in Lebanon, which sought to overturn a pro-Western government and replace it with a pro-Nasser, anti-Western regime.

There was no actual loss of American lives during the rebellion, and the damage to privately owned property of Americans was negligible, although the U.S. Information Agency's reading rooms in Tripoli and in the Basta section of Beirut were looted and burned on 11 and 12 May near the beginning of the uprising. Such rebel leaders as Saeb Salaam and, Kamal Jumblatt strongly declared. that they had no hostile intentions toward Americans or their property. Nevertheless, the statements in the rebel newspapers and over the rebel broadcasting stations were highly derogatory and inflammatory with regard to the United States.[16]

As early as 11 May, Charles Malik, Lebanon's foreign minister, tentatively suggested to U.S. Ambassador Robert McClintock that it might be appropriate to make plans at once for the possible deployment to Lebanon of a division of U.S.

Marines in the event that the intervention of the United. Arab Republic became more overt. McClintock replied that such action might be extremely harmful to Chamoun's position. The United States wished to avoid military intervention in Lebanon, preferring to limit its role to supplying the Lebanese government with a limited quantity of munitions and economic aid. The State Department believed that it would be better for the Chamoun government to sustain itself by its own efforts. Direct U.S. military assistance would provide the Nasserites with the argument that Chamoun's was a discredited government propped up by foreign arms.

The weightlifting ability of the C-124 was a major operational advantage.
Source: U.S. Air Force

At the same time, prudence dictated a certain amount of military precaution in the event of a more serious crisis in Lebanon. On 14 May both the Air Force and. the Navy initiated measures to insure the ability of their forces to deal with any eventuality in the eastern Mediterranean. On 16 May the JCS directed Gen. Lauris Norstad, Commander in Chief, Europe (CINCEUR) and Adm. James L. Holloway, Jr., Commander in chief, U.S. Naval Forces, Eastern Atlantic and. Mediterranean (CINCNELM) to be prepared to airlift one battle group to Lebanon within 24 hours. Tactical Air Command, Strategic Air Command, and. Air Materiel Command were informed of this action. On the 17th, Headquarters USAF ordered TAC to bring 24 fighter aircraft to 24-hour alert for possible deployment to USAFE and directed Military Air Transport Service to send 26 C-124s and 5 support aircraft to USAFE to augment theater airlift. The MATS planes arrived in Germany on 18-20 May.

On 19 May the USAFE airlift planes and the Army battle group were standing by, ready to deploy to the Middle East within 24 hours. On the 22d, U.S. naval units were sufficient for landing operations and evacuation. By then the situation in Lebanon had eased, with Chamoun and. Chehab both indicating that intervention was not desired, and on 23 May the JCS recognized. the relaxation of tension by canceling the alert.[17]

News of the military coup d'etat in Baghdad in the early hours of 14 June resulting in the overthrow of the government of Iraq and the deaths of the principal leaders, reached Beirut almost at once. Although the rebels in the city celebrated the event, which was interpreted as a great victory for Nasser and pan-Arabism, no definite aggressive new moves against the Lebanese government materialized. Whether increased intervention by the Nasserites would, have resulted is impossible to say, since the American landing on 15 July had the effect of freezing the political situation. Ambassador McClintock could find no visible evidence within Lebanon of an increased military threat, and he reported to the State Department his belief that the decision on intervention should be based on political and strategic considerations affecting the entire Middle East.[18]

Nevertheless, President Chamoun asked Ambassador McClintock to call at his office during the forenoon of 14 July and officially requested military aid, from the United. States within 48 hours. Chamoun stated that he was making a similar request of the British and French ambassadors. McClintock found Chamoun highly excited and insistent on the arrival of the Sixth Fleet within the stated time period. By 1248 Lebanon time, the message was on its way to Washington.[19]

Chamoun's appeal for aid was received in Washington that morning (14 July) at 0835 Eastern Daylight Time and remained under consideration for approximately 10 hours. At 1848 hours, following the president's decision to dispatch military aid, the Chief of Naval Operations, as executive agent for the JCS, directed. the commander in chief of the newly activated Specified Command, Middle East (SPECOMME), to execute the operational plan for action in the Middle East, CINCAMBRITFOR 1-58, code name Blue Bat.

This plan for combined Anglo-American operations was to be executed only in part because there was still discussion as to the role that the British would play: whether they would follow the Marines into Beirut or go directly into Jordan, where still another crisis appeared imminent. Since no command headquarters existed at the moment, the Commander, Sixth Fleet, who would furnish a part of the command, was ordered to proceed at best speed to land Marines in Lebanon. Meanwhile, the U.S. Commander in Chief, Europe, and the Commander, Tactical Air Command, were alerted for immediate action in the Middle East. Thus began operation Blue Bat, the first integrated. airborne-amphibious operation to be executed by the United States in peacetime.[20]

In carrying out the Presidential order to give military aid to the government of Lebanon, the JCS ordered five principal military actions. Most of these were already provided for in the Blue Bat plan, but some modifications were made. These actions were as follows:

1. Amphibious landing of the Marine task force beginning at 0900 EDT (1500 Beirut time) on 15 July 1958.

2. Readying one U.S. Army airborne battle group in Europe, capable of airlanding at Beirut airport within 24 hours of an execution order or by airdrop within 36 hours.

3. Preparation for the follow-up airlift of a second battle group from Europe to Lebanon.

4. Deployment of 26 C-124s by MATS to Europe as an augmentation airlift for USAFE

5. Authorization of the substitution of TAC's Composite Air Strike Force Bravo as outlined in TAC Oplan 52-58 (Double Trouble) for the USAFE forces required. in CINCAMBRITFOR Oplan I-58 (Blue Bat).[21]

Thus three major USAF commands would participate in the operation: USAFE would airlift Army paratroopers from Germany to the Middle East; MATS would provide C-124s to augment the USAFE airlift capability; and TAC would provide a counter air strike force as a major combat element of the Specified Command, Middle East. By direction of the president, who feared that a security leak night compromise the success of the operation, only the USAF commands directly involved were informed of the action until after the landing, although the JCS did direct the North American Air Defense Command and the Strategic Air Command to increase their alert posture.[22]

II.   THE AMPHIBIOUS LANDING

CINCAMBRITFOR Oplan 1-58, or Blue Bat, had been prepared during May 1958 when the situation in Lebanon first became serious. At that time, the Joint Chiefs had directed the Commander in Chief, U.S. Naval Forces, Eastern Atlantic and Mediterranean (CINCNELM), to prepare in coordination with United Kingdom representatives a combined operational plan for Anglo-American military intervention in Lebanon and Jordan should it become necessary.[1]

Original Plans Were For Royal Air Force Canberra Bombers To Provide a Substantial Proportion of the Offensive Air Support.
License-built Canberras Served in the U.S.A.F. as the Martin B-57.

Source: Royal Air Force.

Blue Bat called for the employment of two U.S. Army airborne battle groups, reinforced, on station in Germany; the U.S. Navy's Sixth Fleet and an amphibious task force in the eastern Mediterranean; tactical air and transport elements of the U.S. Air Force; and forces of the British army, navy, and air force. The U.S. naval flotilla would include two attack aircraft carriers, and. the British would provide an additional one. The U.S. Air Force would provide one air division headquarters, two fighter-bomber squadrons, one fighter-interceptor element, one composite reconnaissance squadron, one air rescue element, one air refueling element, and medium and heavy transports as required.. The RAF support of the plan included six Canberra, one Meteor, and. two Hunter squadrons, all based on Cyprus and Malta.[2]

The plan provided for initial operations to establish control of the airspace over the area of operations and to secure the Beirut airfield and the rest of the city for use as a base. It was considered. necessary, prior to operations, to obtain authorization to overfly Turkey and to utilize the Adana airbase complex as a staging area; also to overfly, and stage through Libya, France, Italy, and Germany. Later air operations would include the air surveillance of adjacent frontiers in order to control both friendly and unfriendly forces. Fighter escort and. close air support would. be provided as needed for the combined. forces. When the airborne battle groups deployed., they would be airlifted to Beirut and either airdropped or airlanded.[3]

Blue Bat was based on a unilateral U.S. limited. war plan for operations in Lebanon (CINCSPECOMME Oplan 215-58) that had. been prepared in November 1957. The older unilateral plan had been used by the individual USAF commands concerned. as a basis for their own operational plans, but there had been insufficient time by 14 July to convert those to accord with the new combined plan, which was available to them only in outline form. The situation was further complicated by the decision that the British would not participate in Blue Bat -- at least for the time being. CINCSPECOMME accordingly notified his subcommanders on 16 July to delete the word "British" or its abbreviated equivalent from all the command titles in Oplan 1-58 and to follow CINCSPECOMME Oplan 215-58 for detailed unilateral U.S. operations.[4]

USS Plymouth Rock and Amphibious Assault Group
Source: U.S. Navy

On 14 July the force most immediately available to provide the desired military aid in Lebanon was the Second Provisional Marine Force, Fleet Marine Force, Atlantic. Organized in January 1958, this force had moved to the Mediterranean in May, when the rebellion in Lebanon began, to participate in a training maneuver with British Royal Marines and the Italian Navy. It functioned under the Sixth Fleet, being designated Combined Task Force (CTF) 62. It was composed of three transphibrons, each consisting of one battalion landing team, associated equipment, and attack transports and cargo ships. At the time of receipt of orders to land in Lebanon, the nearest of these was Transphibron-6 with Battalion Landing Team (BLT) 2/2 in Area Yankee, south of Cyprus, about 160 miles from Beirut. The other two were considerably more distant: Transphibron-2 with BLT

3/6, en route from Suda Bay, Crete, to Athens, and Transphibron-4 with BLT 1/8, en route from Suda Bay to Gibraltar. But Transphibron-6 was without its landing and beach parties and important equipment because its LSD *Plymouth Rock*, which carried these important elements, was en route to Malta for repairs.[5]

The first indication received. by CTF-62 (the Marine force) of the probability of a landing came on 14 July at 1715 Beirut time in a message from Vice Adm. Charles R, Brown, Commander, Sixth Fleet. Some hours later, at 2350Z, Admiral Brown directed the commander of CTF 61 (the Navy force) to land Marines at Beirut on 11 July, beginning at 1500 Beirut time (0900 EDT). It was estimated that BLT 2/2 would, arrive at H-hour, BLT 3/6 at 0530 on the 16th, and BLT 1/8 at 1800 on the 18th. The Chief of Naval Operations notified. Admiral Brown that it was urgent to make the first landing at the scheduled time and that all possible information on the operation be in the hands of the JCS by 15 July at 0800 EDT, since President Eisenhower was to address the United Nations on that day at about 0100 EDT, presumably to announce and justify the landings.[6]

The Marines of BLT 2/2 went ashore on schedule at 1500 on 15 July in an area south of the city designated. Red Beach. Their air support did not arrive until approximately 15 minutes after H-hour, when seven Navy AD-6s appeared overhead. These had been staged through Cyprus from the CVA *Essex*, which was about 360 miles west of Beirut. Meanwhile naval gunfire support had been available. According to a member of the British training mission, six Lebanese planes were in the air during the

AD-6 Skyraider.
Source: U.S. Navy

landing and had been ordered by the Lebanese G-3 to resist but had refused on the ground that the order had not come through their squadron commander.[7]

If resistance had been encountered it would have been impossible to reinforce the troops ashore for 16.5 hours because BLT 3/5 did not reach the area until 0730 on the 16th. Three tanks for the support of BLT 2/2 landed. on Red Beach over a pontoon causeway at 1602 on 15 July. These had been stored on board an LST that was unable to beach because of the gentle offshore gradient characteristic of the eastern Mediterranean. The absence of the LSD *Plymouth Rock* delayed full logistic support for several days and deprived the Marine unit of two tanks and some artillery.[8]

Slightly more than an hour after landing, at 1510, the Marines entered and secured Beirut Airport. Commercial air traffic was only briefly interrupted., resuming at 1820. It was not until the following day, at 0812, that the naval task force, CTF

61, took control of all air traffic in the area. Meanwhile, unloading difficulties at Red Beach had been eased by the arrival at 2000 on the 15th of the LSD *Fort Snelling* from one of the two transphibrons still en route.[9]

USS Fort Snelling
Source: U.S. Navy

During the afternoon and evening of the 15th, the arrival of Marines within the city of Beirut was anxiously awaited by President Chamoun and Ambassador McClintock, who had received vague information of a possible coup by the Lebanese army and doubted General Chehab's willingness or ability to scotch it. Chamoun, in fact, had concealed from Chebab any information that he had appealed to the United States, and apparently the latter's first intimation of the landing was news of the arrival of the ships. Chebab urged McClintock to dissuade the American commander from disembarking the troops, stating that he feared losing control of the Lebanese army, with subsequent disintegration along religious lines. Chehab stated. "with an air of infinite sadness" that the army had, been the only factor holding the country together and that Lebanon was now doomed. "either to become a Christian Israel or be inundated in the Sea of Islam." McClintock agreed. to pass on to the American commander Chehab's opinion about the landing of troops but warned him that they had arrived to support the established government and that if the army ventured to arrest the President, as was rumored., there would be "most disagreeable consequences." Chebab then promised to "recommend" that the plot against the President be stayed.[10]

The immediate visible reaction to the landings was jubilation among nearly all the Lebanese Christians and. some Moslems. Governmental circles, with the exception of the army leadership and one or two notables, were highly elated. Opposition forces seemed stunned into inaction except for denunciatory articles in their newspapers and. violent statements over their clandestine radio transmitters.[11]

On the night of the 15th the Marines remained at the airport. Ambassador McClintock complained in an angry telegram to Washington that four hours after the landing he was still unable to contact the Navy commander of the landing operations and that his requests and suggestions were coldly ignored by the Marine officer (Lt. Col. Harry A. Hadd) in local command. Apparently the principal matter at issue was the desire of President Chamoun, who still feared a last-minute army coup, for a Marine guard. With tanks at the presidential palace. Chehab had also complained that the Marines had offended the Lebanese military at the airport by telling them to pack up and go home. McClintock requested that immediate instructions be sent to the American commander, still unknown to him, to follow the Ambassador on all political questions.[12]

Admiral Holloway, CINCSPECOMME, arrived with his staff by air at 0400 on the morning of the 16th and set up his headquarters on the USS *Taconic*, which was moored 1,000 yards off Red Beach. His arrival was opportune, as important developments were impending. Transphibron-2 with a second Marine battalion, BLT 3/6, arrived off Red Beach and began landing at 0730 under a precautionary air cover from the carrier *Essex*. Brig. Gen. Sidney S. Wade, commander of the Marine units, also went ashore at 0730 and assumed. personal command of the Marine troops. The operational plan called for BLT 3/6 to move at once to the Beirut Airport and relieve BLT 2/2, which would then enter the city to secure the beaches and the harbor area. BLT 1/8 would be unloaded two days later

Adm. James L Holloway Jnr
Source: U.S. Navy

over Yellow Beach to the northeast of the city to consolidate control of the Beirut area and for possible use in Tripoli, on the coast some 40 miles to the north.[13]

USS Taconic AGC-17.
Source: U.S. Navy

By the morning of the 16th General Chehab had apparently not yet resolved the conflicting points of view within the Lebanese army as to the attitude to be taken toward the Americans. At 0900 he called Ambassador McClintock to protest the proposed movement of American troops into the city and to ask that this be delayed at least until future conferences. McClintock obtained General Wade's consent to a delay and then the two met with President Chamoun at the presidential palace, where Chamoun expressed. his approval of the plan to secure the port area. The Ambassador was recalled alone to a second conference with Chamoun and Chehab a few minutes later, and Chehab again repeated his objections to the Marine deployment in Beirut.

Meanwhile, word had arrived that Lebanese tanks and artillery had taken up a position blocking an advance into the city by the Marine column, which General Wade had now got under way. Chehab, having reluctantly agreed to a passage into the city by the Marines in detachments of three vehicles escorted by Lebanese vehicles, agreed to go with the ambassador to the blockade point. There they found that the Marine column had halted after a Lebanese officer had asked that this be done pending resolution of the higher-level conferences; otherwise his orders were to resist. After a further conference among McClintock, Chehab, and Admiral Holloway, who had now arrived at the scene, Chebab agreed to rescind the orders to his forces. Under additional urging, Chehab's insistence on movement in groups of three vehicles at intervals of 10 minutes also went by the board, and the Marine column moved into the city at 1530 with an automobile containing Ambassador McClintock, Admiral Holloway, and General Chehab at its head. The Lebanese army became increasingly friendly, and the entrance of the Marines was loudly cheered by the local populace.[14]

As the Marines prepared. to move into Beirut on the 16th, ample air cover was available from the aircraft carriers *Essex*, *Wasp*, and *Saratoga* in the waters south of Cyprus. Naval aircraft were scheduled for a fly-over at 0900 of western Jordan, where there were large camps of rabidly pro-Nasser Arab refugees from Palestine. This was postponed at the request of the British so as not to interfere with their movement of paratroops to Amman. Carrier planes from the *Saratoga* made the flyover successfully at the same hour on the following day. This carrier had moved farther down the coast than the other carriers and stood ready to evacuate King Hussein by air from Amman in case of need. [15]

Throughout the Lebanon operation, tactical air support needs, consisting of show-of-force flyovers, leaflet drops, and reconnaissance, continued to be met by carrier-based. aircraft of the Sixth Fleet and, to a lesser extent, by USAF aircraft stationed at Adana, Turkey. The primary employment of tactical aviation was for visual and photographic reconnaissance and. for on-station alert for possible close air support missions. Air operations in the Beirut area followed Navy-Marine Corps control procedures and were under the overall control of the CTF 61 tactical air commander aboard. the USS *Pocono*. The Air Force maintained no air control capability closer than Adana. The only Marine air activity was concentrated in Helicopter Sub-Unit 1, which operated eight helicopters in the immediate area. Embarked on the Wasp, the unit landed at the Beirut Airport on 19 July. Although useful for liaison and reconnaissance, it had a lift capacity of only

32 combat-loaded. Marines. The Marine commander considered that his lack of fixed-wing tactical Marine aviation support would have been a serious weakness in the event of combat.[16]

USS Saratoga off Cannes in 1958. The air group ranged on deck includes A3D Skywarriors, A4D Skyhawks, F3H Demons, F8U Crusaders and AD-6 Skyraiders. Source: U.S. Navy

Transphibron-4 arrived. and began to unload. BLT 1/8 over Yellow Beach northeast of Beirut at 0400 on 18 July. At 1150 on the same day the first airborne troops began to arrive at Beirut. These were 850 men of the 2d Battalion, 6th Marine Regiment, who had been airlifted. from the United States via Port Lyautey, Morocco, in 36 Marine transport aircraft. They moved directly from the airport to ships via Red Beach and then moved to Yellow Beach. where they remained as a reserve force afloat. Unloading of the three transphibrons continued for a total of 7.5 working day's (24 hours each), and 10,397 short tons were put over the Lebanon beaches. During the first three days, slightly more than 6,000 Marines landed in Lebanon, either over the beaches or by airlift.[17]

Altogether, the Marines put ashore 15 M48 medium tanks and 10 Ontos (antitank vehicles mounting six 110 millimeter (sic) recoilless guns). This force was considered. probably inadequate to meet the armored strength of either Lebanon (estimated at 58 tanks) or Syria (estimated. at 207 tanks). There was some slight uneasiness on this score until 27 July, when the U.S. Army landed one tank battalion including 72 M48's plus 17 M79 armored. personnel earriers.[18]

The Ontos Anti-tank Vehicle. Note that contrary to the main text, the guns are 106mm, not 110mm.
Source: U.S. Marine Corps

Meanwhile, one Army battle group and a large support force had been airlifted from Germany to Lebanon. USAFE troop carriers landed 1,749 paratroopers of Task Force Alpha, commanded by Brig. Gen. D. W. Gray, at Beirut Airport on 19 July after a two-day holdover at Adana, Turkey, by order of Admiral Holloway. By 26 July the troop carriers had delivered Task Force Charlie, consistlng of 1,632 men. The second Army battle group, Task Force Bravo, remained on alert in Germany, while Task Force Delta, consisting of 4,411 support personnel, and Task Force Echo, a 90-millimeter gun battalion (558 men), began movement to embarkation ports in France and Germany. The arrival of the last Army units during the first two weeks of August brought the Army forces to a strength of about 8,700 and the total ground force strength, Army and Marine, to about 15,000.[19]

Maj Gen. Paul D Adams
Source: U.S. Army

Because of the size of the above force, which went beyond the Blue Bat concept, Admiral. Holloway requested the Joint Chiefs to assign an Army or Marine Corps two- or three-star general as Commander, American Land Forces. This was approved and. Maj. Gen. Paul. D. Adams, USA, then serving in the European theater, was designated. to the new post on 23 July. Headquarters, U.S. Army Europe was to assist by supplying personnel, equipment, and. headquarters units. General Adams arrived to establish his Joint Headquarters on 24 July.[20]

The Military Sea Transport Service diverted 15 of its cargo vessels to the Lebanon operation during the initial period, and these brought from Bremerhaven alone 3,851 persons and 12,500 long tons of cargo. This included. 88 tanks, 1,906 combat and. general-purpose vehicles, 3 fixed-wing aircraft, and 6 helicopters. These ships were insufficient, however, and soon had to be supplemented by a large number of leased commercial vessels, many of foreign ownership.[21]

By 27 July it was obvious to Admiral Holloway that the saturation point in the Beirut area was near, and he reported to Admiral Burke that no further augmentation was required in addition to Task Forces Delta and. Echo, already embarked and en route. Army Task Force Bravo was thus never called on to move to Lebanon.

The deployment to Lebanon of the Honest John missile battery of Task Force Delta raised questions of far-reaching significance for the U.S. defense establishment. On 22 July, CINCUSAREUR notified CINCSPECOMME that the missile unit would depart Bremerhaven about 24 July and arrive in Beirut about 3 August. He requested concurrence on shipping nuclear warheads by air to arrive one day after the personnel and equipment. This information came as a surprise to Admiral Holloway, who had. Apparently not contemplated the possible use of atomic weapons in the area. He therefore asked that the nuclear warheads be withheld and conventional warheads be sent instead. He was informed, however, that the unit was not equipped with conventional warheads. In view of this situation, Holloway ordered the unit returned to Germany by air immediately upon its arrival.[22]

By 19 July the Navy had assembled. a considerable force in the area immediately adjacent to Beirut. In addition to the 17 ships, which had transported the three amphibious squadrons and their equipment, there were one heavy cruiser, 6 destroyers, and. four minesweepers. At a distance of approximately 150 miles southwest of Beirut -- 33° 33' E -- were 2 attack aircraft carriers, 1 antisubmarine aircraft carrier, 1 guided missile cruiser, 16 destroyers, and 6 destroyer escorts. CINCLANT had also

USS Wasp CVS-18
Source: U.S. Navy

received. orders to be prepared on four hours' notice to steam for Gibraltar. Meanwhile, the Chief of Naval Operations ordered the Commander in Chief, Pacific to load and sail to the Persian Gulf one Marine battalion landing team and to be prepared to augment it. The Navy force in the Persian Gulf consisted of only one small seaplane tender and one destroyer.[23]

During these operations by U.S. forces, other nations were also engaged in military movements as a result of the Lebanon crisis. One French cruiser and three destroyers arrived off Beirut early 17 July in response to the appeal received from President Chamoun. Since American troops had already landed, the French took no further action. While reserving the "right to give orders to land," the French government informed. the United. States that it would not take sudden action. The government of Turkey several times expressed willingness and even eagerness to join the United States and Britain in intervening to suppress the revolt in Iraq. Turkey, however, did not receive any encouragement along this line, and its only military action was to move military units, one cavalry division, to the frontier area near Syria.[24]

The Soviet Union took immediate action following the Iraq revolution to encourage the new government and deter intervention by establishing diplomatic relations and issuing warnings to the Western Powers. It also announced. military maneuvers in Bulgaria, the Black Sea, Transcaucasia, and Turkestan, which seemed designed to dampen any enthusiasm for action by members of the Baghdad Pact. None of these actions or statements was regarded by the United States as particularly threatening; it was considered unlikely that Russia would risk war for its limited stake in the Middle East economy. There was obviously no threat to Russian security involved. It was also pointed out by Air Force intelligence that Soviet diplomatic and propaganda reaction, while prompt and energetic, was not as aggressive as in the Suez crisis of 1956 and that no specific course of action was discernible.[25]

III. DEPLOYMENT OF AIR UNITS

The Air Force was assigned. a major role in four of the five deployments (the first was by Sixth Fleet) ordered by the JCS after the President's decision to send military aid to Lebanon. The Tactical Air Command was to send its composite Air Strike Force (CASF) Bravo to the Middle East. USAFE was to prepare for two separate airlifts of two U. S. Army battle groups from Germany to Lebanon; and in the fourth action, the Military Air Transport Service was to dispatch 26 C-124 transports to augment USAFE's airlift.[1]

On the morning of 11 July, when the first Marine BLT was landing south of Beirut and USAFE transport units were preparing for an airlift to Beirut, the Tactical Air Command was hastily readying CASF Bravo for deployment overseas to Adana, Turkey. Near midnight of the 14th, chief of staff Gen. Thomas D. White had notified Gen. Otto P. Weyland, TAC's commander, that the JCS had approved the substitution of his CASF Bravo force for USAFE units originally scheduled to deploy to Adana in support of Blue Bat. The substitution was being made in order to provide immediate augmentation to USAFE in event of the spread of hostilities. Further instructions would be sent to TAC after 0900 EDT on 15 July. Insofar as possible TAC's troop carriers would replace MATS transports in airlifting CASF Bravo because the MATS planes might be required to airlift the U.S. Army Strategic Army Corps (STRAC). General White instructed Weyland not to alert any units concerning these instructions and to hold the information "closest," but he sent copies of the message to USAFE and MATS.[2]

In the early hours of the 15th, Weyland reported that the following CASF Bravo units were on readiness alert as indicated. in Nineteenth Air Force's Oplan 52-58

2 Sqs	24 F-100D/F tactical fighters	832d Air Division	Cannon AFB NM
1 Sq	12 B-57 tactical bombers	836th Air Division	Langley AFB VA
1 Sq	4 KB-50 tankers	429th Air refueling Sq	Langley AFB VA
1 Sq	Composite recon composed of 6 RF-101, 6 RB-66B, 3 RB-66C, 3 WB-66D	837th Air Div	Shaw AFB SC
	Communications and control element	507th Tactical Control Gp	Shaw AFB SC
	Command element	19th Air Force	Foster AFB TX

Eight tankers were preparing for deployment to Lajes in the Azores, and 24 F-100s would depart from Cannon AFB about 0500 hours for England AFB, La., for further deployment as necessary.[3]

Later in the day, at 1300 hours, TAC notified Headquarters USAF that the units at Cannon AFB could not launch their planes with full loads because of construction work on the runways. To replace them, the 354th Tactical Fighter wing at Myrtle AFB, S.C., would, dispatch 24 F-100D/Fs as soon as possible. The 24 F-100D/Fs from Cannon AFB would proceed to the east coast as a backup force. The three RB-66Cs would not be included. in the reconnaissance group.[4]

A little later, TAC notified Headquarters USAF. that CASF Bravo was being deployed as planned, with the exceptions noted above. The tankers would not deploy to Adana but would remain at route positions. TAC C-130s would provide the support airlift of this force to its destination. The commander of the Ninth Air Force would, control the deployment from his movement control center at Shaw AFB.[5]

The first CASF planes to leave the United States -- the 12 B-57s -- got off from Langley for Lajes beginning at 1420 hours on 15 July. At 1650 hours, the first flight of fighters -- 12 F-100s -- departed Myrtle Beach nonstop for Adana. TAC later reported that 6 RF-101s plus 2 spares, had taken off from Shaw AFB for Chaumont, France, via Lajes, at 1805 hours and that the Nineteenth Air Force command element was en route.

F-100s deployed directly from their home base at Myrtle Beach to Adana in Turkey with the aid of aerial refueling.
Source: U.S. Air Force

The second flight of 9 F-100s departed Myrtle Beach early the next morning, at 0235 hours, 16 July. The 6 RB-66s left Shaw AFB for Chateau-roux via Lajes at 0852 hours that morning, and, the 3 WB-66s got off at 0922 on the same route. A third flight of 8 F-100s departed Myrtle Beach at 1618 hours on the l6th. The first TAC transports had taken off at 1530 on 15 July, and by the end of the day there were 17 C-130s en route to Lajes, moving the command element, the communications and control element, and other support elements as rapidly as possible.

The first 4 F-100s landed at Incirlik Air Base, Adana, at 1125 hours on 16 July after a nonstop flight of 12.5 hours, with three refuelings en route. By the night of the 17th the CASF had 15 F-100s and 10 B-57s in place at Adana. On the 18th, 2 additional B-57s and 6 RF-101s arrived., bringing the total of TAC combat aircraft to 33 (of 54 committed), plus 2 C-130 transports (and 40 others in Europe held en route by CINCSPECOMME because of saturation at Adana). MATS had 36 C-124s at Rhein-Main to augment the USAFE airlift, and USAFE had 58 transports on the Germany-Adana airlift, as well as 9 F-86Ds on air defense at Adana.

On the night of 20 July the CASF at Adana was complete, with 26 F-100s, 12 B-57s, 7 RF-101s, 7 RB-66s and 3 WB-66s. (USAFE's report, Support of the Lebanon operation, gives 60 aircraft, counting 9 F-86s. It lists only 23 F-100s and 6 RB-66s.) Personnel totaled about 1,100. By the morning of the 21st, all of the TAC C-130s, except 3, had departed the Middle East on their return to the United States.[6]

RF-101s Formed The Reconnaissance Element of CASF Bravo
Source: U.S. Air Force

In the meantime, USAFEs 322 Air Division had, airlifted the 1st Airborne Battle Group, 187th Infantry (Force Alpha) from Germany to Lebanon, with a stopover at Incirlik. The USAFE transports began the airlift at 0100 on the morning of 15 July and brought the last element into Incirlik at 1250 hours on the 17th. MATS augmented the USAFE airlift with 10 C-124s that were already in Germany on temporary duty with USAFE and with 26 additional ones that arrived at Rhein-Main on 15-16 July.[7]

On the 18th the 322nd began the airlift of Task Force Charlie from Germany to Adana. Charlie was a support force for Alpha, and it was large enough to support also Task Force Bravo, the second airborne battle group waiting in Germany for orders to deploy by airlift to Lebanon. While some transports were landing Force Charlie at Adana on the 19th, others began the lift of Alpha from Adana to the objective at Beirut Airport, Lebanon. They completed the movement to Beirut just after midnight of 19 July (0039 hours on 20 July).[8]

The key to the Air Force movements and the airlift was Incirlik Air Base, the best military airfield in Turkey, located. a few miles from the city of Adama. Only 80 miles from the Syrian border and about 215 miles north of Beirut, it offered. a base from which aircraft could reach most of the Middle East capitals. The U.S. Air Force leased it from the government of Turkey, but it also served as a Turkish flying training center under Turkish command. Incirlik had a 10.000 foot concrete runway, 21 hardstands, and 6 aprons. Navigational aids, global communications, a weather station, and a floodlighting system were provided, and USAF units were assigned to the base.8 But, Incirlik was still a small base, and saturation of its facilities was inevitable with the simultaneous arrival, beginning on the 15th, of the CASF Bravo aircraft from the United States and the USAFE transports from Germany.

The CASF Components

The Fighters. [9] It was 0910 EDT, 15 July, at Myrtle Beach AFB, S.C., when the commander of the 354th Tactical Fighter Wing received an order to send one squadron of F-100s to Adana, Turkey, within seven hours, followed by a second squadron nine hours later.

The 354ths commander was surprised, for the wing's mission had been changed only on 1 July from that of day fighter to fighter-bomber, and the pilots had no overwater experience and very little night refueling experience. Exercise Tradewind begun just previous to the Lebanon crisis, had shown immediately that the 354th needed considerable unit training in squadron formation and in full-load refueling, both by day and night. Furthermore, the wing had received, its flyaway kits only five days earlier.

According to the Nineteenth Air Force's Oplan 52-58, two squadrons of the 834th Air Division at Cannon AFB, New Mexico., were scheduled as the first tactical units for deployment of a CASF to the Middle East. But, the condition of the runways at Cannon had, made it necessary to substitute the 354ths squadrons.

The 355th Tactical Fighter Squadron was the unit assigned for the first flight deployment, the 352nd Squadron was assigned for the second, and. the third section was a makeup unit composed mostly of pilots whose planes had aborted in the first two flights. Launch hour for the first F-100s was 1618 hours EDT 15 July but actual takeoff occurred at 1650 hours. In all, 29 F-l00s and 37 pilots deployed on direct flight from the United States to Adana. Of the 29 aircraft, 8 were F-100Fs with 2 pilots each, while 21 were of the single-place D series.

A severe operational test en route awaited the pilots of the first squadron of 12 F-100s as they took off from Myrtle Beach late on the afternoon of 15 July. The principal problem was air refueling, and the first test of the skill occurred over the Atlantic southeast of Nova Scotia at 39° 40' N, 61° 51' W. Toward this point, KB-50 tankers had taken off from Langley AFB well before the F-l00s left Myrtle Beach.

The F-100s had been about 30 minutes late in taking off, and they were about 20 minutes late at the first refueling. This placed a serious handicap on the tankers, which had a limited endurance. Only five tankers were on station whereas eight had been scheduled originally. The problems that these F-100 pilots met at the first rendezvous point, as well as those encountered by them and by the other two sections at the two later rendezvous points, included night weather conditions, insufficient number of tankers, inadequate communications, and malfunctioning equipment.

At the first rendezvous, nine F-100s refueled successfully, but three aborted. Two of the three landed at Greenwood, Nova Scotia, but the third ran out of fuel, and the pilot bailed out safely over land, between Yarmouth and Greenwood.

On the flight to the Azores, the aircraft became separated and experienced navigational difficulties. They arrived at the second refueling point in poor weather at night. Five pilots had to land their aircraft at Lajes, but four successfully refueled and continued on to Adana, executing a third. refueling from USAFE tankers over Caritat (near Orange), southern France. The lucky four landed at Incirlik Air Base at 1125 on 16 July after about 12.5 hours in flight.

The second section of nine F-100s deployed more smoothly earlier on the 15th, owing to better preparation. Even so, the number had been reduced. from 12 because one tanker had. aborted and one fighter aborted on takeoff. They carried, through the first and second refuelings successfully but had to land, at Chateauroux when there were no tankers to refuel them on arrival over

Operational Problems Delayed F-100 Flights
Source: U.S. Air Force

Caritat in the afternoon. Eight of these F-100s resumed their flight the next morning and, after one additional refueling stop at Wheelus Air Base in Libya, arrived in Adana on the afternoon of 17 July.

The third section of eight F-100s, composed chiefly of pilots whose planes had aborted earlier, was intended to make up for the aborts of the other two sections and to supply two spares. In spite of pilot fatigue, the takeoffs on the afternoon of the 15th and the initial refuelings were successful. But tanker troubles stopped four of the pilots at Lajes; canopy failure and oxygen shortages caused the remaining four to land at European bases.

The full complement of 26 F-100s was on hand at Incirlik by the 20th. The CASF replaced its 8 F-100Fs with as many F-100Ds near the end of the month. After the Ds arrived, there were 34 Super Sabres at Adana for a few days until the last 6 F-100Fs departed for the States on 25 July.

The Tactical Bombers.[10] The 498th Bombardment Squadron, Tactical (345th Bombardment Wing), furnished. 12 B-57 bombers to the strike force. The first pair of B-57s took off from Langley AFB at 1420 hours on 15 July -- two and a half hours before the first fighters got off. The first C-130 support aircraft for the bombers departed at 1800 hours that evening.

The intended route to Adana was via Lajes and Chateauroux (Deols Air Auxiliary Field, there). Mechanical or communication difficulties forced eight B-57s to land at Ernest Harmon AFB, Newfoundland. Nevertheless, the first B-57 touched down at Incirlik at 0105 on 17 July. On the evening of the 16th, when the first two bombers were a few hours out of Adana, two more were nearing Chateauroux, six were at Lajes, and the last two were en route to Lajes.

B-57s Were The Main Tactical Bomber Force
Source: U.S. Air Force

All 12 B-57s were in place at Adana by 1150 hours of 18 July. The average en route time for the deployment was 40 hours and 7 minutes. The average en route time for the C-130 support aircraft was 47 hours and 55 minutes, the last of these arriving on the evening of 18 July. Among the problems encountered in the deployment was the lack of "Duckbutts" in the flights from Langley to Lajes. The bombers expected at least two such rescue aircraft to be at stationary points en route to facilitate fix positions. In their absence, only airplot information was available for determining geographical location.

The Reconnaissance Planes [11]. The 363d Composite Reconnaissance Squadron, upon arrival at Adana, was composed of 7 RF-101, 7 RB-66B, and 3 WB-66D aircraft. Support personnel and equipment were deployed by TAC C-130s. The personnel of the squadron came from 10 or more units at Shaw AFB, principally the 363rd and 432nd Tactical Reconnaissance Wings and the 837th Air Base Group.

RB-66Bs In Flight
Source: U.S. Air Force

The first of eight RF-101s departed from Shaw for Adana, via the Azores, Chaumont, and Wheelus, at 1800 hours on 15 July. The support aircraft departed that night and early on the 16th and were routed through Bermuda, the Azores, and Chateauroux. The RF-101s refueled in the air northeast of Bermuda. By 1220 hours of 16 July, five of the Voodoos had made successful landings at Chaumont, but three had aborted before reaching Lajes and, returned to Shaw AFB. Two replacements departed from Shaw on the 16th and arrived at Chaumont at 0550 the next morning. Six RF-101s were in place at Adana on 18 July, seven on the 19th.

Six RB-66s and three WB-66s departed from Shaw early on the morning of the 16th. One RB-66 aborted, and returned to Shaw, while two others had refueling difficulties and landed at Kindley, but by 1800 hours of that day, five RB-66s and three WB-66s were at Lajes. These eight reached Chateauroux on the 17th and were directed to remain there because of saturation of the Adana base. They departed, however, for Wheelus on the 18th and arrived at Adana on the afternoon of 19 July. Two more RB-66s arrived on the 20th.

The Fighter-Interceptors [12] The Nineteenth Air Force's plan for CASF Bravo did not include fighter-interceptors because the original deployment plan had called for USAFE to provide them. On 15 July, USAFE therefore offered to deploy eight of its F-86Ds with TAC's CASF. The 512th Fighter-Interceptor Squadron, 86th Fighter-Interceptor Wing, stationed at Sembach Air Base, Germany, was selected for the deployment. This addition to the CASF was ordered on the 16th, and the squadron departed at noon that day. By 0300 hours on 17 July, USAFE had nine F-86Ds in place at Adana, plus two T-33s. The entire force of 69 airmen and 14 officers was on hand. with equipment by 1700 hours. The route had been via Austria, Italy, and Greece, with three refueling stops.

The CASF Support Lift [13] For the airlift of CASF support personnel and equipment, the Ninth Air Force made available the 463rd Troop Carrier Wing (M), Ardmore AFB, Okla., assisted by the 314th Troop Carrier Wing (M), Sewart AFB, Tenn. TAC used 43 C-130s for this lift, although the initial commitment was for 37 aircraft in direct support plus 5 to lift command and en route maintenance elements.

C-130 Aircraft Provided the Airlift For the CASF
Source: U.S. Air Mobility Command Museum

The 463rd received verbal directions to go on Category I Double Trouble alert at 1900 hours (CST) on 14 July. At 1045 (CST) on the 15th, after the aircraft had started taking off for Cannon AFB as directed, the Ninth Air Force advised that the staging base had changed to Myrtle AFB and, at 1100 hours, that the C-130s would be going "all the way." By this hour the number scheduled for deployment was 43.

The transports took on their loads at Myrtle, Shaw and Langley AFBs. The C-130s started taking off for Adana at 1530 hours on 15 July, the first ones arriving there early on 17 July. The route of the C-130s to Adana was through Kindley and Lajes to France (Evreux, Chateauroux, or Chaumont). Although 18 were at Adana on the 19th, most of them were held up at European bases, at the direction of Admiral Holloway, because of saturation of Incirlik facilities when the C-130's began arriving there on 17 July. In some cases the delay in landing was as long as 70 minutes per aircraft. The routing of the entire CASF over the southern route caused, extreme congestion at all terminals, making it necessary to deny the use of Wheelus AFB to the C-130s. The redeployment of C-130s by way of Evreux, either direct or through Wheelus began from Adana on 19 July.

The Airlift of Army Task Force Alpha[14]

On 15 July, while the CASF aircraft were taking off from U.S. bases for Adana, the USAFE troop carriers were preparing to lift Army troops into Lebanon to relieve the Marine amphibious landing teams. The 322d Air Division (Combat Cargo); at Evreux/Fauville Air Base, France, which was charged with the airlift of two Army groups from Germany to Lebanon, had been through a preview of Blue Bat only two months earlier. On 16 May it had received orders to be prepared to airlift within 24 hours troops of the 11th Airborne Division from the Bavarian airfields of Furstenfeldbruck and Erding. By late 18 May, it had 56 transports loaded for airdrop and ready for deployment from these fields. At that time, MATS also sent 31 C-124s from Donaldson AFB, S.C., to Europe as augmentation for USAFE. This crisis in the Middle East abated, however, and the JCS authorized a relaxation in the alert status on 23 May. Following the May emergency the 322d had been placed. on 24-hour, and later 48-hour, alert.

C-124s Did the Heavy Lifting
Source: *Source: U.S. Air Mobility Command Museum*

At 1010 hours on 15 July the 322d received an order to execute its mission as outlined in USAFE Oplan 100-58. A few hours later, USAFE directed the commander of the 322d to airlift Army Force Alpha (the 1st Airborne Battle Group, 187th Infantry. The 187th Infantry was assigned to the 24th Infantry Division.) from Furstenfeldbruck and Erding to Beirut. Departure bases were

Furstenfeldbruck Air Base and Erding Air Station, both near Munich, Germany. It was to be prepared to land or air drop the troops within 24 or 36 hours, respectively, after receipt of orders. The first aircraft were to arrive at Furstenfeldbruck about 1100 hours. The 322d was to assume control of 26 MATS C-124s, plus spares, scheduled to arrive at Rhein-Main Air Base on 15 and 16 July, and it was to set up a joint command post with the 24th Infantry Division.[15]

USAFE had the following planes available in the theater for the Blue Bat airlift mission: 48 C-130s of the 317th Troop Carrier Wing at Evreux; about 50 C-119s of the 60th Troop Carrier Wing at Dreux Air Base, France; some 27 C-123s being phased out of the inventory; and 10 C-124s of the MATS 3d Troop Carrier Squadron stationed at

C-119
Source: U.S. Air Force

Rhein-Main under the operational control of the 322d Air Division. The first 8 C-124s of the MATS augmentation, en route in the theater, were diverted. to Rhein-Main arriving on the l5th, and immediately became available for the airlift. [16]

At 1700 hours on the 15th, USAFE sent a warning order to its task organizations to prepare to deploy for Blue Bat within four hours. About the same time it directed the Seventeenth Air Force, at Wheelus AB, to increase readiness, but covertly, and to prepare for traffic to and from Adana for the CASF and the Army airlift. The commander of the Third Air Force, at South Ruislip, England., was told to double the number of strike aircraft on strip alert.[17]

At 1740 hours USAFE flashed to its units an "alert and movement order" calling for deployment of base support personnel to Adana, airlift for refueling operations and equipment to Germany and Greece, provision of staging capability in Greece and Italy, provision for tactical air control, preparation for deployment of special weapons equipment, and increase in communication services in the Middle East.[18]

For the airlift of Army Force Alpha, USAFE scheduled the transports as follows: 30 C-130s and 5 C-124s would load from Furstenfeldbruck between 1520 and 2010 hours of the 15th; 24 C-119s would load from Erding between 2020 and 2320 hours. Later, the MATS C-124s would replace the C-119s.[19]

By early evening of 15 July the 322d had, 59 transports at the two German fields: 32 C-130s, 8 C-124s, and 19 C-119s. The first C-130 departed with the advance party at 0900 the next morning. Then the troop carriers began the lift of the 1,749 paratroops of Task Force Alpha, the last one clearing for Adana by 1430 hours. The transports completed the lift to Adana in 72 sorties by 1250 hours on the l7th.

Initially, the C-130s flew directly over the Alps from Germany via Austria to Italy and then to Adana, but after Austria protested against the flights, they took the longer route used by the slower C-119s and C-124s, generally via Marseilles,

Naples, and Athens. The principal difficulty resulted from the denial of staging rights in Greece after the 16th. As a result, nine C-119s had to take on additional fuel in Italy in order to fly around Athens, thus diminishing the cabin load and. delaying their arrival at Adana. There were also difficulties at departure airfields until an Air Force component was established to coordinate the loading, dispatch, unloading, and dispersal of aircraft.[20]

By the morning of 17 July, when the situation in Lebanon seemed well in hand to Admiral Holloway, the traffic at Adana was becoming a problem. He asked CINCEUR not to deploy Army Task Force Bravo to Adana but to hold it in Germany on a 24-hour alert until he requested otherwise. Bravo consisted of 1,723 officers and men of the 1st Airborne Battle Group, 503rd Infantry.[21]

On the evening of the 17th, Admiral Holloway directed Army Force Alpha to land at Beirut Airport commencing at 0600 hours of the 19th and to bivouac in the vicinity of the airport. Blue Bat had, required readiness for either an airdrop or airlanding at Beirut, and the initial Army elements had been loaded at Furstenfeldbruck in preparation for an airdrop. Between 16 July and midnight of 18 July, Admiral Holloway conferred with Lebanese officials and ultimately obtained approval for airlanding the U.S. troops at the Beirut Airport. The airdrop loading, the holdover at Incirlik, and. the possibility of a second airdrop at Tripoli complicated the movement.[22]

The troop carriers lifted Task Force Alpha from Adana to Beirut on the 19th, making the first landing at 0549 Beirut time and the last soon after midnight. The Alpha airlift force consisted of 30 C-130s, 7 C-124s, and 19 C-119s, totaling 56 transports. USAFE returned all the C-130s and 9 of the C-119s to the continent from Beirut on the 19th and committed. Them to the lift of Task Force Charlie, after crew rest and maintenance. The remaining 10 C-119s returned to Adana. The C-130s returned. to Evreux, France, via Wheelus AB. 0f the C-124s, 4 were scheduled to return to Rhein-Main, via Naples, but 3 stayed at Adana for shuttle operations.[23]

USAFE had begun the airlift of Task Force Charlie from Germany to Adana at 0200 hours on 16 July. This force was to furnish support for Alpha, and it included field artillery and other heavy equipment. Task Force Charlie was moved simultaneously with Alpha resupply. Together with other resupply, these operations involved approximately 198 C-124 and C-130 sorties, carrying 1.818 Amy personnel and 2,290 tons of cargo. After the first 7 C-124s reached. Adana, the others were delayed until the saturation there could be alleviated. When the Alpha force began departing Adana for Beirut early on the 19th, the lift of Charlie to Adana was resumed. The transports began lifting this force from Adana to Beirut at 0700 on the 20th; the lift was completed, on the 26th of July with the delivery of the 58th Evacuation Hospital in 13 C-124 loads from Rhein-Main. After 20 July the transports operated directly from Germany to Beirut, eliminating the stop at Adana. By 21 July, Blue Bat transports were again authorized to land, at fields in Greece when westbound, but only during darkness when eastbound.[24]

On 26 July, USAFE relocated the units not on the Blue Bat operation to revert to the alert status of 14 July. Beginning 27 July the 322d Division, the MATS planes, could curtail its operations to permit it to perform deferred periodic inspections and to rest its air crews.[25]

In addition to lifting Task Forces Alpha and Charlie to Beirut and providing a fuel lift in response to Jordan's appeal and a supply lift to the British forces in Jordan, the USAFE transports performed continuous support missions for the American forces in the Middle East. By 7 August, they had carried on these support flights 1,269 tons of cargo and 505 passengers. The missions included such special lifts as 125 tons of U.S. Army water pipes to Adana and refueling trucks to Adana for the British.[26]

As the airlift continued, the 322d Division moved a total of 8,227.8 tons of equipment and 7,934 passengers by 12 August. After that, resupply and administrative support continued until the withdrawal of forces between 18 and 25 October. For the withdrawal operation, the division airlifted 1,136.5 tons of freight and, 2,579 passengers back to Furstenfeldbruck.

In its report on the operations, USAFE listed the major problems encountered. First, there was the difficulty in obtaining overflight authority from Austria and Greece. Next, it listed the lack of joint air-ground. operations procedures, not directly affecting the airlift but of great importance.

MATS Airlift [27]

At 1840 hours (CDT) on 14 July, Headquarters MATS received orders to provide 26 C-124s to augment the USAFE airlift. Within three hours, the first C-124 took off from Donaldson AFB, S.C., and thereafter two aircraft departed every hour. The first aircraft arrived at Rhein-Main at 0120 on 15 July, and all 26 C-124s arrived that day, ahead of schedule. In response to a USAFE request on 17 July, the 53d Troop Carrier Wing dispatched an additional 10 C-124s from Donaldson by 2206 on that day. Five replacements were dispatched later.

The night order of 14 July found the 63d Troop Carrier Wing with 21 of its 84 C-124s scattered. over the Pacific, Caribbean, and Atlantic or supporting Army exercises; another 10 were already on temporary duty with USAFE. Only 9 aircraft were on alert status with crew in place at Donaldson. Even for these, changes had to be made to meet requirements for personnel airdrop before they could be dispatched.

Besides the 36 MATS C-124s arriving in the European theater, USAFE had the MATS squadron of 10 C-124s already on temporary duty in the theater, diverted to Rhein-Main AB for the Blue Bat crisis. Seven aircraft of this squadron (the 3d Troop Carrier Squadron, 63d Wing) performed missions in direct support of Blue Bat. After lifting 700 paratroops of Task Force Alpha to Beirut on 19 July, they were assigned to lift POL to Amman, Jordan.

When the crisis occurred on 14 July, the 1607th Air Transport Wing had 13 C-124s en route in the European theater. These were intercepted and directed to Rhein-Main to support the task force; all but one arrived there in the next two days. These aircraft departed for normal duties after the troop carriers arrived from the United States and replaced them. On 15 July, MATS dispatched 3 additional C-124s from Dover AFB to enter the pool at Rhein-Main. The 1607th at Dover had two crews, nicknamed ARAB, held in readiness for departure on one-hour alert at all times. At Charleston AFB, S.C.; and McGuire AFB, N.J., other units were kept on alert.

In mid-August the 150th Wing deployed the 31st Air Transport Squadron with 12 C-124Cs, equipped for troop carrier missions, from Dover to Rhein-Main, where it replaced the 3d Troop Carrier Squadron. Between 13 August and 8 September the 31st flew 44 sorties, most of them to the Middle East.

As of 8 September (the last day of operations) MATS aircraft had moved 5,486 tons of cargo and 5,316 passengers to the Middle East. They had flown 314 sorties in 6,954 flying hours. The operation lasted 56 days, the last four C-124s departing for the United States on 11 September.

The MATS air-base wings at Kindley and Lajes provided refueling and maintenance support to the CASF deployment and redeployment forces as they staged through Bermuda and the Azores. The peak flow occurred on 15 July when 100 military aircraft arrived at Kindley with very little forewarning. On 15-16 July, 49 C-130s arrived within a 24-hour period. 0n 16 July, 161 military aircraft were refueled at Kindley. Lajes serviced more than 300 aircraft during 15-18 July while some 3,000 troops staged through the field.

IV.. AIR OPERATIONS

Operations conducted by the units of all the services during the Lebanon crisis were a "show of force"; there was never any actual combat. Air support over Lebanon came from the carrier-based planes of the Sixth Fleet and, to a lesser extent from the USAF units at Adana. But, since the arena of potential conflict was much greater than Lebanon itself, the Composite Air Strike Force and the USAFE interceptors remained, on the alert over a broad area of the Middle East.

The geographical divisions of responsibility for air defense of the Lebanon area were clearly marked. COMAMARFOR had the mission of defending the northern area, centered on Incirlik Air Base at Adana. This area extended in a semicircle northward around the base into Turkey for about 150 miles, to Turkey's border on the south, then out over the Mediterranean southward to 35° N and westward to a point in the Mediterranean midway between the coast and Cyprus. COMAMNAVFOR had the mission along the Lebanon coast from 35° N to 33° N and out over the Mediterranean to the midway point, where Sixth Fleet took over to the south and the British to the north. The Sixth Fleet kept vigil all the way from COMAMNAVFOR's area to the coasts of, Israel and Egypt. The British area, centering on Cyprus, included all the Mediterranean between 34° N and 36° N beginning at the midway point indicated. The combat reporting center at Adana, set up on 17 July as soon as the F-86s

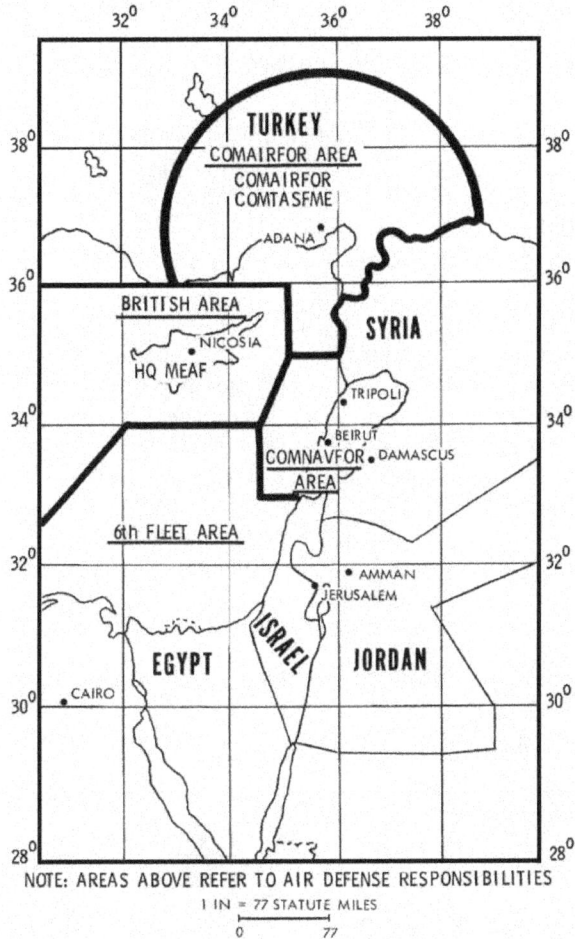

arrived, operated a crisscross telling system with the British forces on Cyprus, and it effected, a tie-in with the Turkish defense net.[1]

Aside from show-of-force missions over Lebanon and the alert and training missions, USAF aircraft flew area cover for Army troop movements to Beirut, executed a leaflet drop mission over Lebanon, and put on air demonstrations for foreign and. national dignitaries at Adana.[2]

The psychological warfare missions were planned and conducted jointly by the U.S. military services and the U.S. Information Service and were controlled by the American ambassador at Beirut. Air Force and naval aircraft periodically conducted flyovers of Lebanon, publicized as "salute to Lebanon" missions. The CASF participated in these operations on 23, 26, 28, and 29 July and 7 and 13 August.

0n 21 July, two escorted. USAF C-130s dropped 1,000,000 leaflets over the populated areas of Lebanon. The message told the people why foreign military forces had entered their country and that these forces would leave when the situation became normal and when requested by the Lebanese government to do so. Wide and favorable publicity was accorded this drop.[3]

Fighters[4]

Four F-100s were in place on the Incirlik strips by 1125 hours of 15 July. At 0800 hours of 17 July they went on a schedule of two on five-minute alert and two on standby. Fifteen Super Sabres had arrived by the 17th, and they flew their initial combat sorties on the 18th. Although only 6 F-100 sorties were flown that day, 16 were flown the following day (19 July) to escort the USAFE airlift from Incirlik to Beirut. By then, 23 F-100s were in place and capable of sustained operations. From that date until the return to the United States of the CASF units on 19 October (except for the 12 F-100s retained in Adana on rotational status), the F-100s were employed as follows:

Type of Mission	Number Sorties Flown
Combat Air Patrols	194
Air Defense	97
Scrambles	205
Flybys	79
Training	141
Aerial Refueling	60
Local Indoctrination	19
Ferry and Administration	59
Engineering	20
TOTAL	**874**

Tactical Bombers[5]

Ten of the 498th Bombardment Squadron's B-57s were combat ready by early 18 July and on 3-minute AOC alert, although fuel pits were not yet in operation. The unit assembled its own rockets and bombs and loaded, five of the planes. Thereafter the highest priority was placed on armed standby alert.

The B-57s flew missions as follows: day and night visual reconnaissance; formation show of force; high and low-altitude navigation training; gunnery, rocketry, and napalm training; firepower demonstration; close air support training; logistic support; courier, administrative, instrument, and test flights. The B-57s participated in show-of-force formations on 23, 26, and 29 July.

During the period of operations, an in-commission rate of 87 percent was maintained for the B-57s. One B-57 received minor damage from gunfire, probably .30 caliber, while flying a visual reconnaissance sortie over Lebanon on 9 August.

The need for additional training in the use of conventional weapons was soon recognized. Many aircrews were untrained in the low-level delivery methods for conventional bombs, and few had fired guns or rockets for score. In early September an air-to-air range over the Mediterranean Sea became available for training in splash gunnery with conventional ordnance. From 22 September through 3 October the bombers used to good advantage for scored gunnery, rocketry, and napalm drops the air-to-ground range at Eskisehir, Turkey, about 240 miles northwest of Incirlik.

Air Defense [6]

The F-86 crews of the 512th Fighter-Interceptor Squadron from USAFE had the mission of air defense of the forces at Incirlik Air Base during hours of darkness. They assumed the first alert commitment 30 minutes after sunset on 17 July -- the day of their arrival. Remaining on the operation for 97 subsequent days, they flew 506 sorties, which included air defense and training missions. Besides night alerts, the squadron augmented the day fighters at times and assumed the commitment during weather below the minimums for day fighter operations.

Reconnaissance [7]

For the CASF as a whole, reconnaissance bore by far the heaviest portion of the operations because of the Army requests. The entire 363rd Composite Reconnaissance Squadron was in place at Incirlik Air Base on 19 July and ready for operations at 1900 hours, but missions did not begin until 21 July. These missions, generally routine and well within the capability of the unit, included weather, day photo, and visual reconnaissance, mostly low-level. The RB-66Bs did not fly night missions because of possible misunderstanding by the inhabitants of night illuminants and. flash equipment. Weather reconnaissance missions were flown by the WB-66s until 20 August, but there was little need for them because of the prevailing fair weather.

Some doubt arose as to the effectiveness of the RB-66B for visual reconnaissance. For good vision, it had to be flown too low and too slowly to avoid being an excellent target for ground fire, and on four missions the RB-66s were hit by ground fire not larger than .30 caliber. The RF-101s were used effectively.

Although one of these was also hit by ground fire, the RF-101s were of course less vulnerable because of their smaller size and greater speed.

Since aerial reconnaissance made up a major portion of the operations and both Air Force and naval aircraft were used by the Army for this purpose, problems arose concerning common procedures for requesting and reporting reconnaissance. Also there was need for a base map that all the services could use. When this problem became evident, representatives of the services met in Beirut on 4 August, devised joint procedures for directing and controlling air operations, and crafted an interim aerial reconnaissance plan.

The major deficiency in operations resulted from the deletion of the electronic reconnaissance aircraft (RB-66C), which had been in the plan at the beginning of the deployment. The CASF could not meet a requirement for precise information on radar capabilities within Syria because it did not have the RB-66Cs. USAFE supplied one Q-54 sortie for this purpose. The Nineteenth Air Force commander agreed that three RB-66Cs always should be included in a CASF.

Tankers [8]

No tankers were scheduled to deploy with the CASF to the Middle East; the tankers used, for refueling the aircraft on the initial. deployment remained at route positions. Later, however, on 20 July, after the CASF was in place at Adana, four KB-50J tankers of the 429th Air Refueling Squadron deployed there through Lajes and Chateauroux, arriving on the 21st.

From 21 July through 7 September, the four tankers were on one-hour alert, the aircrews sleeping beneath their planes at first and later in nearby tents. They performed air refueling missions for training purposes and also flew transport missions. After two tankers returned to the United States on 7 September the other two remained on two-hour alert. The tanker operations, in general, showed. the importance of close coordination between receiver and. tanker units in all planning, briefing, and. critiques.

Airlift to Jordan

The situation in Jordan, to which military aid was given by the United Kingdom in close coordination with the United States, differed in almost every respect from that in Lebanon except that both countries had been under heavy pressure from Nasser and his followers and the governments of both felt themselves directly menaced by the violent revolution in Iraq. Jordan closely resembled the other Arab countries in that the dominant religious group was Sunni Moslem with other groups constituting only a small minority. Unlike Lebanon and some other Middle East countries, it had only a short national history, being an artificial country set up with the support of the British in 1920, in large part as a reward for the support of the Hashemite family against the Turks during World. War I. For more than 30 years the British continued to pay an annual subsidy to Jordan and, also detailed military officers to aid in organizing an army. As a result, Jordan was the only Arab country to emerge with any credit from the war with Israel, actually gaining

38

new territory and population at the cost of additional economic and political problems.

Following the assassination of King Abdullah in 1951, the throne ultimately passed to his grandson, Hussein, who assumed power in 1953. Three years later, Hussein dismissed Lt. Gen. John B. Glubb, the British officer who had built the Jordanian army into an effective fighting force. A few months afterwards, during the Suez crisis, Jordan broke off the long-standing alliance with Britain and renounced its annual subsidy in return for promised financial assistance from Egypt and. Saudi Arabia. This help failed to materialize, and a series of political crises, military conspiracies, and mob disorders instigated by extreme Arab nationalists and Communists, racked the kingdom. Only repeated purgings of conspirators from the officer corps, the loyalty of the army rank and file, and the support of his cousin, Faisal of Iraq, had enabled Hussein to hold on to his kingdom. The coup in Iraq on 14 July threatened to dislodge him from his throne. [9]

Two days after President Chamoun's appeal for aid, at 1100 local time on 16 July, Hussein appealed to the United States and the United Kingdom for military assistance. There is every indication that Hussein would have preferred that the United States share in the Jordan movement, but the two governments decided. that the British alone should send troops into Jordan. Since the Lebanon operation was already under way, however, and the U.S. carrier task force was much closer to the scene, the commander of the Sixth Fleet did receive contingent instructions to prepare to evacuate King Hussein. The carrier *Saratoga* was sent farther down the eastern Mediterranean coast, and plans were made to use two planes plus air cover for the evacuation, if it became necessary. [10]

The Blackburn Beverley provided British heavy airlift. Source: Royal Air Force

On receipt of Hussein's appeal and with the concurrence of Jordan, the British immediately requested. and obtained permission from the government of Israel to overfly the country, the only stipulation being that Israel be notified of the timing of the flights. The nearest British force of any size capable of land operations was about 250 miles north-west of Amman, on Cyprus, where the 15th Parachute Brigade and 1st Guards Brigade were located. Early on the morning of 17 July, 500 British paratroopers left Cyprus aboard. 14 Hastings and 8 Beverley transports. By 2122 local time they were taking up defensive positions outside the Amman airport. Six Hawker Hunters from No. 66 Squadron on Cyprus provided fighter cover. [11]

The first phase of the British movement into Jordan was completed by 21 July, when about 1,800 troops were on hand (the 16th Parachute Brigade less one battalion). The number of British troops in Jordan ultimately rose to 3,500. The British made precautionary deployment of additional forces in North Africa and the Arabian peninsula. Headquarters 3d Infantry Division was airlifted from Britain to Cyprus, while Headquarters 24th Brigade began movement to Bahrain, where 1,000 British troops were already stationed. Several hundred additional British Marines landed in Tobruk, bringing the number of British troops in Libya to 3,500. The lst Guards Brigade remained on alert in Cyprus, for movement to Jordan or Kuwait as might be necessary. A number of British troops also noticed to Aden.[12]

The serious shortage of POL in Amman from the beginning prompted a hurried. appeal to the United States for aid. The nearest source of supply was at Bahrain, and shipment was first planned in C-124s via the American air base at Dhahran, Saudi Arabia. The latter country, however, refused to give overflight clearance. On a long-term basis, the problem was dealt with by the shipment of fuel by ship tanker through the Suez Canal to Aqaba, without challenge from the Egyptian government. But the immediate need on 17 July was pressing and could be solved only by airlift.

CINCSPECOMME met the crisis by flying in fuel from Beirut. In this first U.S. lift to Amman, completed on 26 July, 7 C-124s carried 361 tons of fuel, and 13 C-130s, diverted from the support of Task Force Charlie, lifted 247 tons. Israel permitted. the planes to fly over its territory but specified a minimum altitude of 14,500 feet, which precluded the use of C-119s, considered more efficient on short runs.

In the meantime, the British had requested another airlift of supplies from Cyprus to their forces in Amman. This airlift, to run, from 24 July to 1 August, was needed because land transportation from Aqaba to Amman was totally inadequate. The United States agreed to this assistance on 21 July, and. Israel notified Secretary of State Dulles on the same day that U.S. planes had clearance to overfly Israel.

For this lift CINCSPECOMME requested 9 C-119s to add to the aircraft already being used on the Amman POL lift. Subsequently, USAFE provided 16 C-119s for the Jordan lift. (Smart memo of 1 August says 23 C-119's.) The transports picked up empty drums in Beirut, hauled them to Nicosia on Cyprus, and filled them up there for the lift to Amman. More than 1,600 empty drums were found in Beirut and, with 23 rubber collapsible fuel containers, these were sufficient for the lift. By 31 July, the U.S. transports had. airlifted 1,572 tons of cargo to Amman.

On 28 July the British also requested a daily airlift of 102 tons of cargo for an indefinite period after 1 August. On 27 July, the British had in Amman a 21-day supply of stocks and a 37-day supply of POL. The JCS authorized Admiral Holloway to assist the British on a temporary basis to 5 August but advised that there were not enough transports to permit a continuing airlift. CINCSPECOMME expected the overflight agreement with Israel to extend. through this period, but on 2 August the Israeli government denied overflight rights and operations were suspended.. The lift was resumed on 6 August when Israel again granted clearance.

On 1 August, 6 C-124s arrived in Cyprus to take over the haul to Amman from the C-119s and continue the resupply through 10 August. By 6 August the British in Amman had sufficient POL supply for 57 days. Counting the initial lift, the U.S. transports carried a total of 2,277 tons of cargo to Amman.[13]

The Command of Air Units.

The command arrangements for the American forces evolved substantially in accordance with plans as the operation in Lebanon unfolded. The Commander American Air Forces (COMAMAIRFOR) and the naval and ground commanders (COMAMNAVFOR and COMAMLANFOR) took their orders from Admiral Holloway, the overall commander of the operation (CINCSPECOMME). But the title COMAMAIRFOR did not carry with it the command of air units belonging to the U.S. Navy, although COMAMLANFOR embraced all ground troops after 24 July. The air commander had only TAC's CASF Bravo directly under him and exercised operational control of the USAFE units when they arrived in Turkey and Lebanon (the area east of 28° E). He was also the coordinator with the British Middle East Air Force and the Turkish Air Force.

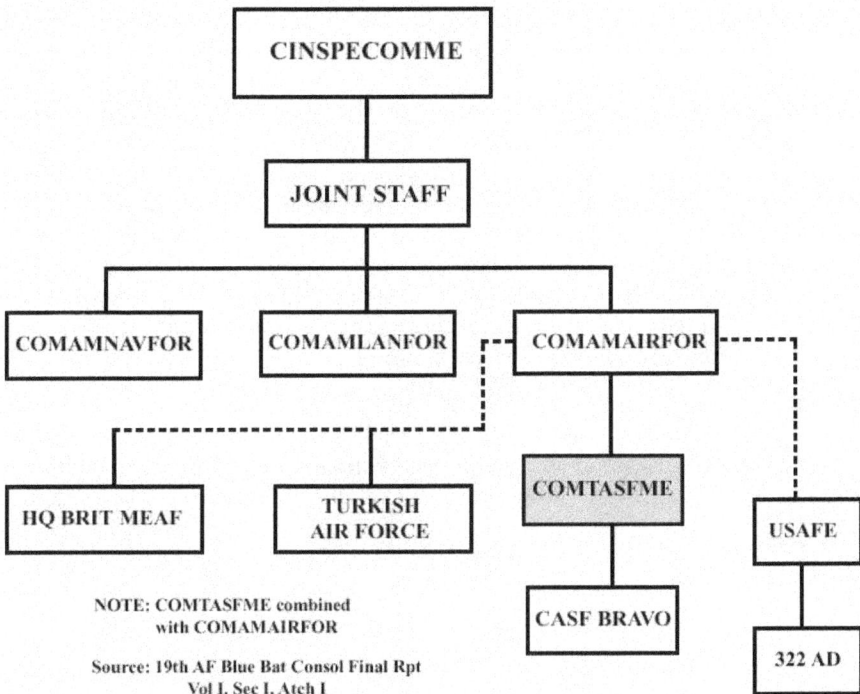

NOTE: COMTASFME combined
with COMAMAIRFOR

Source: 19th AF Blue Bat Consol Final Rpt
Vol I, Sec I, Atch I

An advance staff for the air commander established itself at Adana by 1000 hours of 16 July and the TAC CASF command element arrived there at 0800 hours the next morning. On 21 July, Admiral Holloway called for the activation of the land commander's headquarters ashore in Beirut and directed. that advance elements of the air commander's staff be located adjacent to the former.[14]

At the beginning of the Blue Bat operation, Brig. Gen. James E. Roberts, from USAFE headquarters, held the position of COMAMAIRFOR, but Maj. Gen. Henry Viccellio, commander of the Nineteenth Air Force and CASF Bravo, replaced him on 21 July. Soon afterward the advance element of a small air headquarters was established in Beirut near the land force headquarters.[15]

The need for standardized joint doctrine and procedures for air missions became apparent as soon as the Air Force brought the Army troops to Beirut. Since joint use was made of Air Force and naval aviation for aerial reconnaissance, close support, air defense, and other air operations, it was necessary to coordinate requests for missions and to develop procedures that would be satisfactory to all the services and could be coordinated with the RAF on Cyprus. At a conference on 4 August, joint procedures were devised and an interim aerial reconnaissance plan was drafted. According to Admiral Holloway's report, this proved adequate for the purpose, but future plans would provide for joint control and coordination of all aspects of air operations. [16]

COMAMAIRFOR and. COMAMNAVFOR (Rear Adm. Charles R. Brown) agreed upon a plan whereby the air support responsibilities in Lebanon would be rotated between them. On 5 September, General Viccellio assumed this duty from the naval commander and on 22 September it was rotated again. On 27 September the responsibility for control of air operations was shifted from CTF 61 on the *Pocono* to COMAMAIRFOR. Admiral Holloway informed the two sub-commanders that General Viccellio would assume from Admiral Brown the complete responsibility for tactical and transport air operations in the Lebanon area, This meant that Viccellio would provide air defense warning to the land forces; maintain active air defense alert at Adana; maintain two aircraft on 30-minute AOC alert for close air support and the remainder of the aircraft on 2-hour alert; provide reconnaissance and training aircraft as agreed with COMAMLANFOR assume full responsibility for search and rescue in the Lebanon area and provide required communications.[17]

Cooperation between the American and British forces brought appreciative words from the commander of the Middle East Air Force on Cyprus. RAF Air Marshal Sir Hubert Patch discussed air coordination with Admiral Holloway at Beirut and with General Viccellio at Adana. Patch spoke of the splendid U.S. cooperation in the airlift to Jordan, so essential to the success of the British forces there. [18]

V.. CONCILIATION AND WITHDRAWAL

During their stay in Lebanon the American forces remained prepared to protect Beirut from attack, but they took no aggressive action against rebel forces inside or outside the city. In general, the Americans were posted in the vicinity of the airport to the south of the city, at Yellow Beach to the northeast of the city, and along the harbor frontage. All units stayed on the alert and. conducted patrols, and Air Force and Navy units executed a number of low-level flyovers.

The ground forces were to fire only if fired upon. Then they could return fire, employing the next larger weapons if available. No significant actions resulted although several minor exchanges of fire occurred. Desultory firing from the ground against low-flying American aircraft occurred on a number of occasions, and several planes were hit, but no significant damage or casualties resulted. Actually, the only serious American casualty caused by hostile rebel action was an unarmed soldier, who was shot and wounded in the left shoulder on 23 August while walking near a barricade outside the Basta section of Beirut. In response to Admiral Holloway's vigorous protest, General Chehab had the barricade destroyed.[1]

Lebanese who supported Chamoun, as well as those who feared the destruction of life and property that might result from a continuation of the rebellion, greeted the newly arrived American troops with relief and enthusiasm. This included virtually all the Christians, except for a few politicians leagued with the rebels, and a certain number of Moslems. According to Admiral Holloway, this feeling was maintained and even increased during the operation. The good deportment of the troops and the improvement of business resulting from the more stabilized situation and from the presence of the Americans were doubtless important factors.[2]

The American command was concerned about concluding a status of forces agreement with the Lebanese government, but for various reasons, particularly the greater interest of the Chamoun government in more pressing matters, this was delayed for some time. The Lebanese foreign ministry finally consented. on 6 August to a status of forces agreement, effective on the 6th, that gave the Americans the rights and exemptions normally provided by such arrangements.[3]

Although the rebel radio and newspapers violently denounced the Americans, after the first few days the rebel leaders and their organized forces became increasingly circumspect in their attitude. A few American patrols that wandered into rebel-held territories were captured and disarmed, but the members were quickly released together with their vehicles. The rebels became less truculent as the operation progressed, although some of the clandestine radio transmitters continued to be intensely belligerent. It was reported that on 20 July rebel leaders had decided to "avoid activities that might enable Chamoun [to] provoke fighting between U.S. forces and [the] opposition." On the 22d, Saeb Salaam, chief rebel leader in the Basta, issued a press and radio statement cautioning his forces not to fire on American troops, and on the 24th he sent word to the American ambassador that he had no hostility toward Americans or the troops that had arrived in Lebanon.4

The arrival of the Americans had stimulated intense political activity by all sides to find a compromise solution of Lebanon's internal problems. Under Secretary of State Robert Murphy, who arrived from Washington on 17 July, joined with Admiral Holloway and Ambassador McClintock during the first days of the intervention in urging President Chamoun and General Chehab to drive the rebels out of their stronghold in Beirut -- the Basta. Chamoun favored the operation because he desired a military success to restore his prestige prior to leaving office; he also claimed. that it would enhance Chehab's. Holloway was unwilling to use the American forces for aggressive action, but he offered to take over guard duties that would free Lebanese units for action. Chehab, while not positively refusing, managed to delay and evade the issue. Because he was necessary to a successful compromise, he could not be dealt with in summary fashion, although Holloway and McClintock seriously considered it at one time.

Murphy quickly became convinced that only a political settlement was practicable, and he and McClintock conferred with most of the rebel chiefs and secured indications of their willingness to compromise. The proposal to attack the Basta was allowed to drift until finally dropped.. As time passed, Holloway and McClintock became increasingly friendly to Chehab and less so to Chamoun. Eventually, both became convinced of Chehab's wisdom and patriotism and spoke of him in the highest terms.5

By 23 July, Chamoun, while postponing the election from 24 to 31 July, indicated that he was ready to accept Chehab as his successor, as were almost all the other leading figures on both sides. The election was almost postponed again as the result of an unsuccessful attempt on the life of Prime Minister Sami Solh on 30 July, which Solh blamed on Chehab.5

The election of Chehab on 31 July did not have an immediately perceptible effect on the situation in Lebanon. The rebels insisted that Chamoun resign immediately after the election, but he declined to do so. As a result the period between 31 July and the inauguration on 23 September was one of uneasy waiting, with barricades remaining up and the general strike continuing. Chehab remained as head of the army.

Meanwhile, on 21 August the possibility of stabilizing the situation in the Middle East was greatly enhanced by the passage in the United Nations Assembly of a resolution unanimously approved by the members of the Arab League. Without going into details, the resolution was a compromise declaration made possible by a distinct retreat from the extreme stand formerly taken by the United Arab Republic.[7]

During early September, despite a second attempt on the life of Prime Minister Sami Solh, the outlook for a settlement remained favorable. Meanwhile, there was much political activity in preparation for the change of administrations, with both Chehab and the American leaders involved in repeated conferences with the leaders of the opposition to Chamoun. Dissension and even fighting broke out among the rebels over the question of accepting a compromise, but it was significant that a steady stream of Syrian "volunteers" began to leave the country. As the date of the inauguration approached it became obvious that Chehab's choice as the new prime minister would, be Rashid Karame, Moslem leader from Tripoli, who was considered the most moderate of the opposition. It was worth noting, however, that even he had publicly stated that he regarded Nasser as a Superman.[8]

On 19 September, when military activities seemed to be giving way to political in preparation for Chehab's inauguration on the 23d, the trend was interrupted by the kidnapping and, presumed murder of Fouad Haddah, an employee of the Phalangist newspaper, who had written critically of Nasser. Wild disorder swept over Beirut as a result, with the pro-Chamoun Phalange carrying out reprisal attacks and erecting barricades that sealed off the Asrafiya, the Christian quarter of the city, and blocked many of the main roads leading into Beirut. Although the inauguration of Chehab took place without incident and the general strike came to an end, the Phalange called a new general strike of indefinite duration. Contrary to expectation, it was generally successful outside the purely Moslem section.

Rashid Karame
Source: Lebanese Gov.

The Phalange gained additional support following Chehab's appointment of Karame as prime minister, the latter's ill-timed remarked about "reaping the fruits of the revolution" and the appointment of a cabinet made up entirely of former oppositionists (revolutionists) plus two Third Force adherents. The situation was now completely reversed, with the former opposition supporting the government and much of the former government support behind the barricades. The possibility of a religious war seemed to hang in the balance for a short time, but the disorders quickly subsided into a new stalemate, with the Phalange demanding the punishment of Haddah's murderers, the resignation of Karame, and representation of the Chamoun faction in the goverrment.[9]

Several factors made a settlement possible. Both sides were tired of the struggle and, once the momentary excitement passed, were willing to discuss a compromise. The American forces doubtless helped restrain all-out warfare, and, the Lebanese army took increasingly vigorous action in suppressing disorder. It also was obvious that Karame's one-sided cabinet could not command a majority in the national assembly and, to avert the turmoil of a new election, would have to provide more equitable representation of the political factions, particularly the powerful Chamoun bloc. With Ambassador McClintock acting as an intermediary among Gemayel, the Phalangist leader, Karame, and Chehab, a formula of "peace without victory" was agreed on, and a new and more broadly representative cabinet, though still headed by Karame, was installed on 14 October. There immediately followed the calling off of the Phalange's general strike, a removal of barricades, and a return to normalcy by 20 October. The army attempted to carry out a collection of weapons held by civilians, but most were probably stored for later possible use. Chehab later confided to the American ambassador that the strike of the Christians had been a good thing because it showed the Moslems that mutual concessions were necessary.[10]

Withdrawal of Ground Forces

The American movement into Lebanon, as has been seen, occurred not so much because of the situation within the disturbed country itself as the possibility that the sudden and. violent revolution within Iraq might engulf, the other Middle Eastern countries. Within a few days it became obvious that the upsurge of violence would be confined to Iraq. On 25 July, Headquarters USAF notified all concerned commands that the peak of the crisis had passed and that they could reduce the alert status to permit necessary combat crew training.[11]

The situation continued to improve, and by 5 August the JCS directed Admiral Holloway to submit recommendations concerning reductions in his forces and to begin planning for an orderly but prompt withdrawal of troops and materiel following Chehab's assumption of the presidency. Pressure for a token withdrawal of part of the force at an early date had come from the Department of State, which believed it politically desirable in order to influence world opinion. After a conference with Ambassador McClintock and General Chehab, Admiral Holloway agreed that a token withdrawal would have a beneficial effect within Lebanon, since it would deprive the opposition elements of one of their chief propaganda weapons. Chehab had altered hid original opposition to the presence of American forces to suggest that the bulk of the forces be left intact until peace was assured in Lebanon.[12]

Accordingly, Holloway requested. discretionary authority to withdraw at least one and possibly two battalions of Marines with maximum publicity during the following week. After further discussions with Chehab, the withdrawal movement of Marine Battalion Landing Team 2/2 began at 0700 on 13 August and was completed two days later. The withdrawal was loudly criticized by most Christian elements of the population, but it was received. with satisfaction by the opposition. [13]

The main phase of the withdrawal got under way on 15 September, when other Marine units, principally BLT 1/8, were embarked. On the following day the first Army unit, the 299th Engineer Battalion, began loading its heavy equipment on board the USS *Lieutenant Craig*. The last Marine battalion, BLT 3/6, was loaded on 28-30 September despite the renewed disturbances within the country. This left only U. S. Army forces in Lebanon, primarily the airborne task force. Marine BLT 3/6 remained in the vicinity at sea and was available for relanding in an emergency.[14]

The withdrawal continued during most of October as shipping became available, with heavy equipment preceding the tactical units. Supplies and equipment of selected logistical elements were loaded beginning 6 October, tanks of the 3d

Medium Tank Battalion beginning 12 October, personnel of the same battalion and of other units on 15 October, and the sea-tail of the airborne troops on the same day. The airlift of Army troops began on 19 October and was completed on the 25th, when the Commander American Land Forces, General Adams, also departed. Admiral Holloway had shifted his headquarters to London two days earlier after a farewell reception given by President Chehab.[15]

Earlier, on 16 September, Admiral Holloway had proposed that a U.S. Army training mission of 11 officers and 60 men be left behind to reorganize the Lebanese army which he described as ill prepared to cope with internal or external contingencies. He further proposed that a small American force of 1,200 men be left behind. This would be supported by a USAF tactical air squadron at Adana, by two destroyers of the Sixth Fleet on patrol in the Beirut area, and by one transphibron with a Marine BLT embarked no farther west than 20^0 E longitude. Although Ambassador McClintock concurred in this recommendation, it apparently did not receive favorable consideration. [16]

Withdrawal of Air Forces

The first element of the CASF to redeploy to the United States was the weather reconnaissance unit. The three WB-66s departed for the United States on 23 August, followed by the RB-66s on 5-6 September, leaving six RF-101s to handle reconnaissance activities. The fighters began their redeployment on 4 September, the first section of eight F-100s departing Incirlik that day for Myrtle Beach via Wheelus and. Chateauroux. A second section of eight F-100s took off for home on 13 September. Also in early September, CINCUSAFE released the remaining MATS augmentation aircraft, and the last C-124 departed for the United States on 11 September.[17]

The 322nd Air Division set up an airlift task force headquarters at the Beirut Airport on 15 October to coordinate the evacuation of airborne troops and certain ground personnel. For the return, permission was secured to overfly Austria, and an arrangement was made for notifying the Austrian government of the exact penetration time. The airlift of Army Task Force Alpha began at 0001 on 19 October. Most of the C-130s, which comprised the majority of the aircraft, returned to Furstenfeldbruck by flying over the Austrian Alps after stopping at Brindisi, Italy, for refueling, although some went by way of Marseilles. The C-124s returned to Furstenfeldbruck by way of France, after stopping at Naples for refueling. When this airlift ended, on 25 October the transports had lifted 1,136.5 tons of cargo and 2,579 passengers in 77 C-130 and 13 C-124 flights. [18]

The remaining CASF units began their departure from the Middle East on 19 October -- 6 F-100s,12 B-57s, and 6 RF-101s leaving on that day. The F-86s of USAFE's 512th Fighter-Interceptor Squadron flew back to Sembach Air Base in Germany on 21 October.

On 16 October the JCS had given temporary permission to the Air Force to retain one F-100 rotational squadron at Adana after the withdrawal, but a final decision would be subject to the agreement of the Turkish government. The control of the squadron would pass to CINCUSAFE who would delegate it to the Seventeenth

Air Force. Twelve F-100Ds of the 353d Tactical Fighter Squadron arrived at Incirlik to replace the departing F-100s.

On 24 October the air headquarters was deactivated along with the land and naval headquarters. With the conclusion of the airlift back to Germany on the 25th, the Blue Bat operations came to an end.[19]

The WB-66s Led The Way Home
Source: U.S. Air Force

VI .. CRITIQUE AND CONCLUSIONS

Although the Lebanon crisis resulted in no combat action by the American forces, it constituted. the most considerable emergency movement of U.S. forces since the Korean War and the first ever undertaken by the country in the Middle East proper. The rapid movement of troops and aircraft over thousands of miles in response to a critical local situation required widespread coordinated activities by the three military services in planning, operations, intelligence, and logistics. These activities were complicated by intricate diplomatic and political considerations. Under these circumstances it would have been remarkable if conclusions useful for the future had not been derived.

At the higher reaches of government, on 9 February 1959 the JCS approved for submission to the chairman of the National Security Council Planning Board a report that listed. Five principal lessons of the Lebanon operation as most suitable for consideration by the NSC.[1]

1. The need for early determination of overflight and staging rights.

2. The need for early action to secure statue of forces agreements.

3. The need to inform the American public quickly on reasons for intervention.

4. The need for adequate facilities to avoid congestion and delay during limited war operations that require rapid, deployment of forces, equipment, and. supplies by air.

5. The need for a budget plan to cover joint emergency operations.

Most of the problems encountered by the Air Force resulted from the lack of adequate facilities and procedures to meet either scheduled or unscheduled requirements. Some problems also grew out of the abbreviated warning time received by units. For units previously scheduled. to participate in the operation, the warning time was not unreasonable, but for the unscheduled and substitute units, it was inadequate. In large measure the overtaxing of already strained facilities resulted from changes in movement caused by the use of substitute and additional units.

The CASF Bravo fighters and. TAC's C-130 transports were the chief victims of changes in the assignment of units to the operation. Two squadrons of the 354th Tactical Fighter Wing substituted for squadrons of the 832d Air Division with only seven hours' notice, had difficulties in getting underway, especially the first flight. The last-minute changes were confusing, the time for preparation was too short, and the shortages were legion—maps, let-down books for foreign bases, radar facility charts, exposure suits, and communication information.

TAC's C-130s received short warning notice because the plans had named MATS to be responsible for airlifting the CASF support force, and it was only late on 14 July that TAC was told to use its own C-130s for this purpose. And on the morning of 15 July the signals were changed again when the main staging base was changed from Cannon AFB, N. Mex., to Myrtle Beach AFB, S. C. The C-

130s were not told until 1100 hours (CST) on 11 July that they would have to go all the way to Europe and Turkey. They started taking off from the United States for Adana only two and a half hours later.

The C-130 crews underwent the hardship of performing straight-through flights without autopilots to the Middle East after having had their aircraft diverted from missions in progress. The tight time schedule, plus the inadequacy of loading plans and the complete absence of standardized loading procedures for the Hercules, caused confusion at the staging bases in the United States. Much on- and off-loading resulted when individual crews enforced different load maximums and cargo distributions. The C-130 en route maintenance crews found it almost impossible to get adequate rest during flight because of the size of the load and the noise of the engines. The crews did a good job, but fatigue was a serious deterrent to effectiveness.[2]

MATS, too, felt that the warning time had been inadequate. Although it had been scheduled, to provide C-124s to augment the USAFE airlift, MATS believed that with more strategic warning time, it could have reduced its reaction time. When it received deployment orders, the C-124s were scattered far and wide, and it took 34 hours to get the first 26 to Germany from the United States.

USAF bases involved in the operation also suffered from the lack of adequate warning and the changes in plans. The MATS bases at Kindley and Lajes were not warned by either MATS or TAC that TAC C-130s would replace C-124s in the support lift of CASF Bravo. Accordingly, they were not prepared to support the movement as effectively as would have been desirable. Wheelus AFB in Libya found itself, unexpectedly overwhelmed by aircraft arrivals because the denial of overflight or staging rights by Austria and Greece made it necessary for many of the planes to fly by the southern route instead of more directly from bases in Germany and France.[3]

But these bases had simple problems by comparison with those of the Incirlik base at Adana. This was the only American base in the Middle East that could be used for the Lebanon operation, and it was quickly saturated with men and planes. The facilities at Incirlik proved inadequate to support the forces assembled there, and operations suffered accordingly.

Most of the units reported similar experiences. The lack of housing was especially acute, and the crew of the 353d Composite Reconnaissance Squadron and of the tankers from the 129th Air Refueling Squadron had to sleep on the ground until tents or some kind of housing could be made available to them. Other types of shelter – those for operations and maintenance -- also were severely limited.. Utilities at Incirlik quickly proved inadequate, and the shortage of water was especially critical. Ground transportation and landline communications could not carry the load, nor could the radar for ground control of interceptors meet the needs of the 512th Fighter-Interceptor Squadron from USAEE. Supply shortages included AOCP items and photographic items. Part of the trouble arose from the inadequacy of the flyaway kits the planes brought with them -- especially the tankers and the reconnaissance planes.[4]

The fundamental problem underlying the operational and logistical difficulties was the lack of bases in the operational area. Incirlik alone could not have supported a USAF force engaged in combat operations over Lebanon and other areas of the Middle East. The lack of operational bases is a problem that may well arise in other areas of the world where the United States has commitments. The problem should be given the most serious consideration in planning for emergency situations.

The difficulties encountered by the Air Force during this operation stemmed also from deficiencies in the key operations plan – CINCSPECOMME Oplan 215-58. There was no provision in the plan for control and coordination of joint air operations, and the procedures had to be worked out after the forces arrived. The absence of such procedures would have had grave consequences if combat operations had begun immediately on the arrival of the air units. The lack of more precise command arrangements in the plan could also have had serious effects had actual combat been necessary.

Administrative and logistical arrangements were slighted in Oplan 215-58, and the effects were felt by Air Force units, especially at Incirlik. The medical annex did not specify a representative of one of the services to be the surgeon on Admiral Holloway's staff; this omission resulted in a lack of coordination among the services that would have been most serious had there been combat operations. Much of the confusion of loading and unloading at air bases also could have been avoided had. provision been made ahead of time for Joint Army-Air Force control groups at the principal airfields.[5]

Another oversight in the plans was the failure to provide for the assignment to the landing forces of officers who could speak Arabic and had some knowledge of Lebanon. Nor was there provision for rapid communication between the American ambassador and the landing forces. Apparently the Navy and Marine commanders of the initial landing were not instructed on the role of the ambassador with respect to their operations. This led to misunderstandings and irritations that could have had more fateful consequences had the landings been made under less favorable circumstances.

In the broadest sense, apart from the consideration of specific errors in performance or planning, the 1958 Lebanon operation may possibly be best considered under three headings: Was it necessary? What was its effect? Did the organization and composition of the force and the movements executed represent the most economical use of forces? None of these basic questions is considered specifically in any of the official military studies of the operation, although most of them assume or imply that the operation was a success because the situation became stabilized. in both Lebanon and. Jordan. The necessity for the operation is not questioned, probably because it was the result of a politico-diplomatic decision rather than a military one. The question of force composition and tactics, although likewise hypothetical, is on the other hand obviously a matter for military analysis and decisions and might well have been examined at JCS level. The general conclusion in the JCS report of 19 April that the operation was conducted in an orderly and efficient manner hardly disposes of this matter.

The Lebanon

The U.S. decision to send a military force to Lebanon was apparently based. largely on the assumptions that the unexpected revolution in Iraq was directly stimulated. by Nasser and that it could very possibly result in violent repercussions in Lebanon and Jordan by UAR military intervention or by direct instigation or sympathetic reaction of local Nasserites. The American ambassadors in Beirut and Amman did not believe that the danger to the existing governments was appreciably increased by the Iraqi developments. On the other hand., the chief executives of both Lebanon and Jordan were sufficiently alarmed to request military aid, and the United States granted it under the policy established in the Eisenhower Declaration of 1956.

It now appears that American intelligence was defective both in its failure to forecast the revolution in Iraq, which was considered to have a more stable government than either Lebanon or Jordan, and in analyzing the revolution's nature and sources after the event. Today it is clear that the events in Iraq, though supported by Nasser, were quite beyond his control, and. an actual purge of the Nasserites followed within a few months.

From the above it might appear that in terms of forestalling the seizure of the governments of Lebanon and Jordan by Nasser supporters the military action in both cases was unnecessary. In a wider sense, however, it may be logically argued that it was desirable and necessary to give encouragement to friendly governments in the Middle East and to display a readiness and ability to deploy prompt and. adequate military aid. This was particularly important with respect to the remaining three Asian nations of the Baghdad Pact -- Turkey, Iran, and Pakistan -- as well as to the countries considered to be directly menaced – Lebanon and Jordan -- and to such vacillating Arab countries as Saudi Arabia Sudan and Libya.

Although Ambassador McClintock at first doubted the need for the American landing in Lebanon, he became convinced before the withdrawal that it had been highly effective, as did Admiral Holloway. Both emphasized that the operation exerted a calming influence on the warring factions and enabled the negotiation of a compromise settlement with a minimum of further bloodshed. They were gratified that no casualties of consequence resulted from the operation and that the leaders of the task force parted on excellent terms with the new heads of the government as well as with the officer corps of the Lebanese Army. The last two contentions were easily demonstrable and represented no mean achievements.

The success in calming the rival factions is less demonstrable, since fighting continued on some scale after the arrival of the Americans and even intensified in late September. Although the American military used direct force only in a few brief exchanges of small arms fire, the presence of the units clearly had its effect. On several occasions the threat of possible American military action was raised in discussions by Ambassador McClintock with leaders of recalcitrant factions, and in each instance the threat was effective.[6]

On the question of whether the composition of the military force and its movements were of optimal nature, Gen. Maxwell D. Taylor, U.S. Army Chief of

Staff at the time, stated that the circumstances of the operation were so favorable that it was not prudent to draw general conclusions from it, technically successful as it was. He mentioned the advance planning permitted by adequate warning, the limited force required, and the absence of combat operations. To these factors might be added the convenient geographical situation, which permitted quick access by both sea and air, and the favorable attitude of at least half the population together with the government officials of the country. In other words, the character of the operation was not such as to create any particular strain on the existing force structure of the United States for limited war.

Conveniently available ground forces proved sufficient for the operation, all units coming from the European and Mediterranean areas except for the 850 Marines flown from the United States and kept afloat as a reserve. The United States easily met Chamoun's condition that help arrive within 48 hours; the first Marines landed 24.5 hours after the request for aid was received in Washington. To do this, however, Admiral Brown had to send. Forward ahead of his main force a single battalion landing team of 1,600 men. Only two tanks were put ashore with this force, and the very limited air cover did not appear until 15 minutes after the first landing. Furthermore, it would not have been possible to reinforce significantly the first landing group for at least 15 hours. The full Marine force of 6,000 men was not unloaded until three days after the first landing, and even then included only 15 tanks. By then, more adequate air support was available, since aircraft from three carriers and from the Air Force units at Adana (29 combat planes) were available. Air Force transports landed the first contingent of Army troops -- more than 1,700 men -- on 19 July, but no Army tanks arrived until 27 July.

Actually, of course, U.S. ground, naval, and air forces turned out to be more than adequate. The area was cut off by the *cordon sanitaire* of the Baghdad. Pact nations from direct contact with the Soviet Union and its satellites and only the Syrian portion of the UAR was adjacent to Lebanon. It is even possible that, short of Russian intervention, the American force could have dealt with any organized armed force that might have been mustered against it in the immediate vicinity, although the occupation of large areas would have undoubtedly required many more troops. Had further prompt extensive reinforcements been required, beyond the limited number on alert in Germany, they could have come only from the United States by air transport.

The Nineteenth Air Force report on Blue Bat pointed out that the operation had. been of the "lightest requirement." The objective area – Beirut -- was only a little more than 200 miles from the base – Adana – and only a small force was deployed from the United States. The problems and deficiencies that did arise would have been greatly magnified by a graver situation.[7]

Any extension of operations undoubtedly would have called for additional tactical air support and air transport, placing a heavy burden on the Air Force. The air support could have come at first only from the one crowded base at Adana. The Air Force was in the position of a backstop for operations in which the Army and more particularly, the Navy were playing the leading roles. Although the Air Force underwent no great test, what would have been the effect had the Taiwan

Strait crisis occurred simultaneously with the Lebanon crisis instead of later? Certainly, both TAC and MATS might have found their resources either inadequate or seriously strained. Probably, the Air Force would have had to draw on long-range transports earmarked for the support of the Strategic Air Command or to conscript them from civilian airlines, which would have required the declaration of a national emergency. MATS tried to contract with the civilian airlines for additional lift during the Lebanon crisis, but the airlines would not be persuaded of the necessity without declaration of a national emergency. [8]

PART TWO – AIR OPERATIONS IN TAIWAN

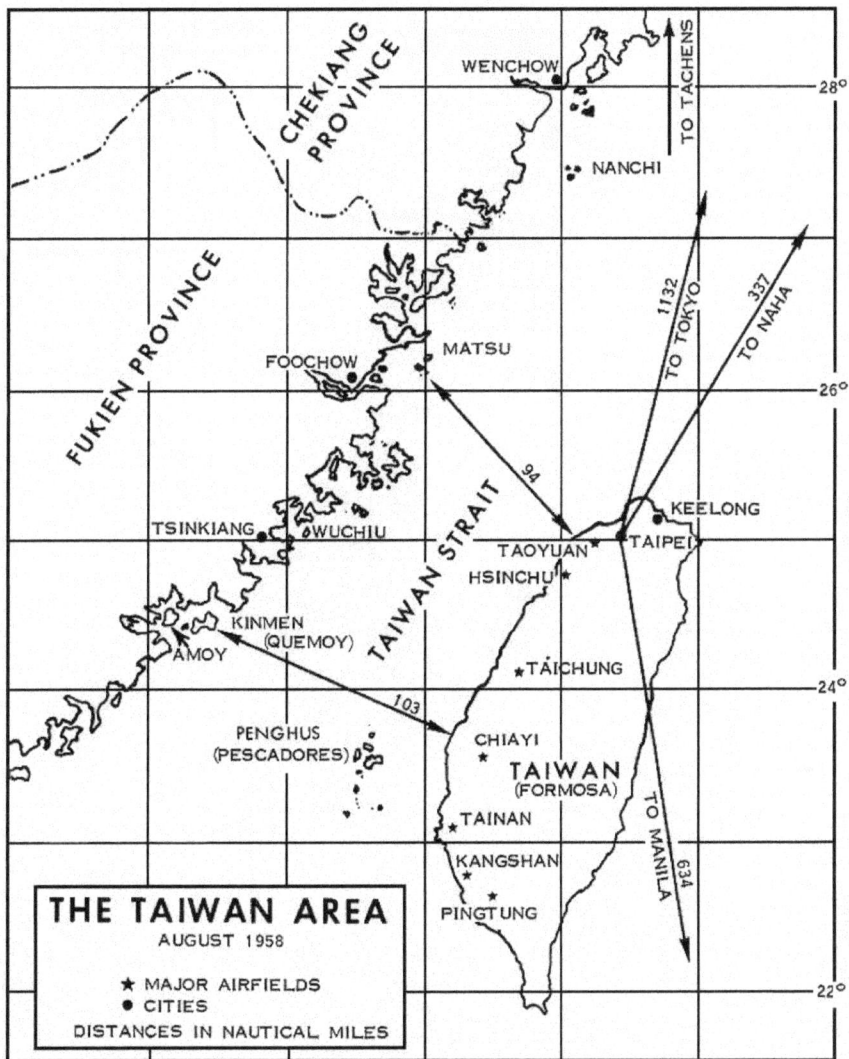

CHEKIANG PROVINCE

WENCHOW

NANCHI

TO TACHENS

28°

FUKIEN PROVINCE

1132 TO TOKYO

337 TO NAHA

FOOCHOW

MATSU

26°

TSINKIANG WUCHIU

TAIWAN STRAIT

94

KEELONG

TAOYUAN TAIPEI

HSINCHU

KINMEN (QUEMOY)

AMOY

★TAICHUNG

24°

103

PENGHUS (PESCADORES)

CHIAYI

TAIWAN (FORMOSA)

TO MANILA

634

TAINAN

KANGSHAN

THE TAIWAN AREA
AUGUST 1958

PINGTUNG

★ MAJOR AIRFIELDS
● CITIES
DISTANCES IN NAUTICAL MILES

22°

0 _____ 46   1 INCH' – 46 STATUTE MILES

I. THE TAIWAN CRISIS OF 1951 – 55

Lying about 100 miles off the mainland of China is the island of Taiwan, also known to the Western world as Formosa. Its 13,429 square miles roughly equal the combined area of the states of Massachusetts, Connecticut, and Rhode Island. Taiwan is 243 miles long and from 50 to 80 miles wide. With the U.S.-held island of Okinawa 350 miles to the northeast and the Philippine Islands 225 miles to the southeast, Taiwan is one of the most important military bastions in the western Pacific. About 25 miles from the island, in the Taiwan Strait, are the Penghus, a duster of islets also called the Pescadores. Historically an appendage of Taiwan, they too are important because of their strategic position.

The people of Taiwan are primarily of Chinese stock, derived mostly from Fukien and Kwangtung provinces, although some come from the south China plateau. About 2 percent are aborigines. During the twentieth century the population of Taiwan expanded rapidly, from about 3.5 million in 1920 to about 10 million in 1956. The latter figure included some 2 million refugees who fled from the mainland in 1946 - 49. About 480,099 Japanese were repatriated to Japan in 1945.

China ceded Taiwan and the Penghus to Japan at the end of the Sino-Japanese War of 1894-95, but 50 years later these territories were again in Chinese hands. In the Cairo Declaration of November 1943, the United States together with the United Kingdom and the Republic of China called for the return to China of all territories taken from that country by the Japanese. At the Potsdam Conference in July 1945 the United States and the United Kingdom, with the concurrence of China and the subsequent adherence of the Soviet Union, signed a declaration stipulating that the terms of the Cairo Declaration be carried out. With the defeat of Japan, China assumed administrative control over Taiwan and the Penghus. In September 1951, as a last formality, Japan and. 26 other nations (not including the Soviet Union) signed a treaty of peace wherein Japan renounced all title and claim to these islands.

At the end of World War II the civil war that had been smoldering since the late 1920's broke out anew between the Chinese Nationalist government under resident Chiang Kai-shek and Chinese Communist forces entrenched in north China. By March 1948 the Communists controlled Manchuria; by year's end they had seized virtually all of the country north of the Yangtze River. Defeat followed defeat until the Nationalists were driven off the mainland of China. They fled to Taiwan, the Penghus, and a number of offshore islands, primarily the Kinmen (Quemoy), Matsu, and Tachen groups. The Kinmen group consists of Big and Little Kinmen plus four other all within 10 miles of Fukien Province. Big Kinmen, about 13 miles long and 8 miles across at its widest point, lies only five miles from Fukien Province and communist-held Amoy Island. The Matsu group consists of seven islands also about 10 miles from Fukien Province. they lie just outside the Min River estuary at the northern entrance to the Taiwan strait and are about 155 nautical miles northeast of the Kinmens. The Tachen group consists of

2 main islands, Shang and Hsia, about 10 miles from Chekiang Province, plus 23 smaller islands. The Tachens lie about 250 nautical miles north of Taiwan.

In December 1949, Taipei became the provisional capital of the remnants of the Republic of China. Meanwhile, in Peiping on 1 October 1949, the Chinese Communists established a new People's Republic of China, a regime recognized by the Soviet Union two days later.

U.S. Policy Toward Taiwan. 1949 – 54

General George C Marshall
Source: United States Army

During the immediate postwar period a United States mission, headed by Gen. George C. Marshall, made an unsuccessful attempt to mediate the Chinese civil war. After the Nationalists lost mainland China, the U.S. State Department attributed the defeat to internal political, economic, and military weaknesses that had been beyond the power of the United States to remedy save through outright intervention. Although sympathetic to the Nationalists, the United States initially avoided a policy that would risk involvement in the Chinese civil war. President Truman stated on 5 January 1950 that the U.S. Government would provide no military aid for Chinese forces and desired no special rights, privileges, or military bases on Taiwan.[1]

The outbreak of the Korean War on 25 June 1950 ended this policy and led to the military neutralization of Taiwan. Observing that its loss to the Communists would threaten the security of the Pacific area and U.S. forces serving in that area, President Truman directed the U.S. Seventh Fleet to prevent an attack on Taiwan. At the same time, the fleet was ordered to prevent any sea or air operations by the Nationalist government against the mainland. In July, the JCS affirmed the strategic value of the island to the United States. Their recommendations resulted in renewal of military assistance to the Nationalists in February 1951 and the establishment of a Military Assistance Advisory Group (MAAC) in Taipei in November 1952.[2]

In February 1953, shortly after assuming office, President Eisenhower altered U.S. policy toward Taiwan when he declared that the U.S. Seventh Fleet would cease to shield Communist China from the Nationalists.[3] However, the "unleashed" military forces of the Chiang government did not attack the mainland, and the action had no major impact on the Korean War.

The cessation of hostilities in Korea in July 1953 eased tension in the Far East only temporarily. The Chinese Communists renewed their assistance to the Indochinese revolutionaries who were fighting French rule. French defeats led to the establishment of a Communist government in North Vietnam. This prompted the United States to enter into new military defense arrangements. It took the lead in concluding a seven-nation Southeast Asia Collective Defense Treaty, signed in Manila in September 1954., Then, on 2 December, it signed a bilateral mutual defense treaty with the Chinese Nationalist government, assuring the defense of Taiwan and the Penghus and such "other territories" as might be determined by mutual consent -- an allusion to the small, offshore islands.[4]

In an exchange of notes between the two governments on 10 December 1954, the United States recognized that the Nationalists possessed the inherent right of self-defense not only for Taiwan and the Penghus but also for "other territory" under their control (i.e., the offshore islands). Or their part the Nationalists accepted a limitation on their freedom of action. The use of force in this area by either the Nationalist government or the United States would be a joint decision except in an emergency when the Nationalists clearly might have to defend themselves. The notes reflected U.S. fear that conflict with the Communists might arise from unilateral action by the Nationalists who fervently wished to repossess the mainland.

By this time, the status of the Nationalist-held offshore islands had become an increasingly acute problem for the United States. The Communists made unsuccessful attempts to invade the Tachens in May and August 1954, Big Kinmen in September, and Wuchiu Island, about 60 miles northeast of Big Kinmen, in November. In addition, the Communists on the mainland engaged in periodic artillery duels with the Nationalists on the Kinmens and Tachens. Against this background of tension, U.S. Secretary of State John Foster Dulles, on 1 December 1954, publicly discussed the U.S. commitment to defend the "other territories" referred to in the mutual defense treaty. Explaining that the status of the offshore islands was not affected by the treaty, he said that their defense by the United States would depend on the nature of the specific islands under attack and

John Foster Dulles
Source: United States Congress

whether this attack was deemed part of the defense of Taiwan. On this matter the President would probably make the final decision.[6]

U.S. Air Operations During The 1954 – 55 Crisis

Meanwhile, Chinese Communist forays against the offshore islands and limited air action by both the Nationalist and Communist air forces spurred U.S. military preparations in the area. The JCS had transferred responsibility for the defense of Taiwan, the Penghus, and the Philippines from CINCFE to CINCPIC on 15 March

1952. On 4 September 1954, the JCS warned the Commander-in-Chief Far East (CINCFE), Gen. John E. Hull, in Tokyo, and the Commander-in-Chief Pacific (CINCPAC), Adm. Felix B. Stump, in Hawaii, that it might be necessary to implement the latter's Ops Plan, 51-53. This plan called for augmentation of CINCPAC's air and naval units and for U.S. participation with Nationalist forces, if authorized, in the defense of Taiwan, the Penghus, and some of the offshore islands. In approving the plan the JCS had directed the establishment of a USAF component of the Pacific Command, and, on 1 July 1954, the Air Force had formed the Pacific Air Force primarily for this purpose. Its commander was subordinate to the commander of the Far East Air Forces (FEAF) -- a major command of CINCFE -- in matters pertaining solely to the Air Force and subordinate to CINCPAC in matters relating to the defense of Taiwan, the Penghus, and the Philippines.

In November, JCS directed the Air Force to designate a fighter wing within FEAF for dispatch to Taiwan on short notice. For FEAF, the possibility of deploying a wing posed manifold logistical, maintenance, communication, and fuel problems. FEAF was also mindful of CINCFE's concern lest the transfer of too many military units to Taiwan jeopardize the defense of other areas (Korea, Japan, and Okinawa). It was finally determined that a show of force, if necessary, could be made from Taiwan by rotating squadrons of FEAF's 18th Fighter-Bomber Wing (F-86s) from Japan to the island as part of a training and familiarization program. The State and Defense Departments concurred in the rotation plan on 22 January 1955 during another period of rising tension in the Taiwan Strait.[7]

On 17 January, Ichiang, a small island northwest of the Tachens, had fallen to communist forces. On 24 January, President Eisenhower asked Congress for a resolution to authorize him to employ the. armed. forces promptly and effectively to assure the security of Taiwan and the Penghus. The President's request was approved by overwhelming majorities in the House and Senate late in January. The resolution passed by the Congress read as follows:[8]

President Dwight D Eisenhower
Source: U.S. Government

Resolved by the Senate and House of Representatives of the United States of America in Congress assembled. That the President of the United States be and he hereby is authorized to employ the Armed Forces of the United States as he sees necessary for the specific purpose of securing and protecting Formosa and the Pescadores against armed attack, this authority to include the securing and protection of such related positions and territories of that area now in friendly hands and the taking of such other measures as he judges to be required or appropriate in assuming the defense of Formosa and the Pescadores. This resolution shall expire when the President shall determine that the peace and security of the area is reasonably assured by international conditions created by action of the United Nations or otherwise, and shall so report to the Congress.

Also on 24 January, JCS, with State and Defense approval, ordered the entire 18th Fighter-Bomber Wing to Taiwan. the change in plans resulted from a sudden U.S. decision to evacuate about 40,000 Chinese Nationalist soldiers and civilians from the Tachen Islands, which were considered indefensible. Ships and amphibious elements of CINCFE's naval forces joined the U.S. Seventh Fleet in assisting the evacuation. Following preliminary air deployments on 26 January, three squadrons of the 18th wing flew from Japan and the Philippines to bases at Chiayi, Tainan, and Taoyuan on Taiwan. Transports of the Fifth Air Force's 315th Air Division lifted personnel, supplies, and equipment.[9]

To direct air operations from Taiwan, the Fifth Air Force established Headquarters, Air Task Force Fifth (Provisional) at Taipei. Its mission was to cover the Navy's evacuation of the islands, protect the carrier *Princeton*, and, in coordination with the Chinese Nationalist Air Force (CNAF), provide for air defense, search, and rescue in an area south and west of a line 27°N 125°E to 28°N 121°E8.[10]

On 3 February the 18th Wing had on hand 55 combat-ready aircraft and 78 combat-ready pilots. For a brief period it flew training and familiarization missions. During the evacuation of the Tachens, which began on 9 February and continued until the 13th, the 18th flew 206 sorties, 184 of them over the Tachen and Nanchi Islands in direct support missions. On 24 February the Nationalists had also voluntarily abandoned the Nanchi group, midway between Taiwan and the Tachens and 25 miles from the China coast. The CNAF, jointly responsible with the U.S. Navy for antisubmarine patrol south of 27°N, flew P4Y aircraft at 1,000 feet and maintained top cover with their F-47s and F-84s. All forces were cautioned to remain at least three miles from the Chinese mainland. In accordance with JCS instructions, operational control of the 18th Wing passed from CINCFE (FEAF) to CINCPAC (Pacific Air Force).[11]

Thus, what had been initially planned as a show of force through a training exercise became a tactical operation. After the successful completion of the evacuation without Communist opposition, the Taiwan Strait crisis began to wane. By the end of February the 18th's aircraft returned to their home stations, ending CINCPAC's control over the wing. The United States continued to display its air strength on Taiwan, however. CINCPAC's squadron rotation plan was put into effect by JCS directive. The first unit to rotate, the 69th Fighter-Bomber squadron, temporarily stationed at Dark AB, flew to Chiayi AB in February 1955. Other fighter-bomber or fighter-interceptor squadrons followed, initially at about two-week intervals and after 1 July 1955 for longer periods.12 In January 1958 units again began to deploy more frequently for shorter periods.

The Tachen evacuation focused attention on the ability of U.S. and Chinese Nationalist forces to defend Taiwan, the Penghus, and the off-shore islands and to conduct joint air operations from Taiwan. The operational problems were formidable. The major deficiency was communications, for U.S. forces had to rely on inadequate and unreliable Nationalist on- and off-island telephone, teletype, radio, and radar systems. The Joint Operations Center (JRC), a Chinese agency, was almost inoperable, and the use of Chinese installers, operators, and maintenance men created a language barrier that compounded communication difficulties. An important logistical deficiency was the severe shortage of F-86 drop tanks-sufficient for only two days of sustained operations. An intricate U.S. command structure on Taiwan complicated both logistical support and liaison with the Nationalists. There was also an urgent need for closer U.S. and Nationalist defense planning and the CNAF needed more and better aircraft. [13]

To rectify some of the deficiencies, plans were prepared for improved communication and radar systems and air-base facilities. A U.S. section of the Chinese JOC was established in June 1955, Meanwhile, U.S. and Chinese Nationalist representatives held a series of defense coordination conferences in March and April to follow those held in May and December 1953 and October 1954. As a result of the latter conference, Admiral Stump established a Formosa Liaison Center in May as an integral part of the Formosa Defense Command, permitting more intimate U.S. and Chinese working relationships in planning operations, and training. The CNAF was also strengthened by the acquisition of F-86F fighter-interceptors. Contrary to U.S. military advice, the Nationalists bolstered the defenses of the Quemoy group with an additional division of troops from Taiwan. [14]

Interval Between Crises

Nearly three and a half years elapsed before Chinese Communist pressure against the offshore islands created a new crisis in August 1958. During this interval, the United States and the Nationalists made numerous changes in their defenses in the area. The U.S. Air Force off-set a decrease of units and personnel by introducing more modern aircraft, capable of carrying a variety of nuclear weapons, and by stationing a Matador missile squadron on Taiwan in February 1958. The expansion of the CNAF also helped to offset the decrease in USAF's numerical strength. FEAF was redesignated the Pacific Air Forces (PACAF) on 1 July 1957 and the Pacific Air Force was dissolved at that time. By the middle of 1958, PACAF possessed 879 aircraft, including 129 bombers and 401 fighters, The CNAF had nearly 500 jet aircraft, largely F-84s and F-86s. As it grew, air facilities on Taiwan improved, but many inadequacies remained. The communication net was especially weak, since U.S. plans for major expansion of the net had been rejected as too costly. [15]

The period also witnessed the development of a new role for the Tactical Air Command (TAC). In 1955, Headquarters USAF directed TAC to develop a mobile Composite Air Strike Force (CASF) with an atomic capability, to be used in small localized wars. Earmarked for CASF under TAC's plan Blue Blade (subsequently renamed Double Trouble) were fighter, fighter-bomber, reconnaissance, troop carrier, refueling, communication, and supply units. Assigned to the Nineteenth Air Force for specific operations, CASF units made practice deployments, including one in November 1957 to the Far East under the code name Mobile Zebra, A PACAF-TAC agreement signed in May 1958 set forth the conditions under which TAC forces would deploy to the Pacific. [16]

The U.S. command structure in the Pacific also underwent change. The U.S. Navy emerged as the executive agent for the entire area on 1 July 1955 when the Far East Command with headquarters in Tokyo was disestablished and its responsibilities were transferred to the Pacific Command in Hawaii. In the realignment of subordinate commands, also with headquarters in Hawaii. PACAF was the principal USAF command in the entire Pacific area.[17]

At JCS direction, service commanders sought ways to further streamline arrangements on Taiwan. Here three separate headquarters existed: Taiwan Defense Command (TDC), Military Assistance Advisory Group (MAAG), and Air Task Force (ATF) 13 (P). The Air Force and the Army wanted to eliminate TDC and consolidate all military activities on Taiwan under a Chief, MAAG, who would be an Army lieutenant general, but Admiral Stump considered such an arrangement impractical and politically infeasible at the time. He cautioned against a change that might imply a decline in U.S. interest in Taiwan. Early in 1958, with JCS approval, TDC and MAAG were combined into a single headquarters under a vice admiral. The commander of ATF 13 also served as chief of Air Force Section MAAG. This centralization of command provided one rather than three points of contact between U.S. and Chinese military authorities and also permitted reduction of the U.S. headquarters staff and facilities on Taiwan.[18]

U.S. AND CHINESE NATIONALIST COMMAND RELATIONS IN THE TAIWAN AREA

12 MAY 1958

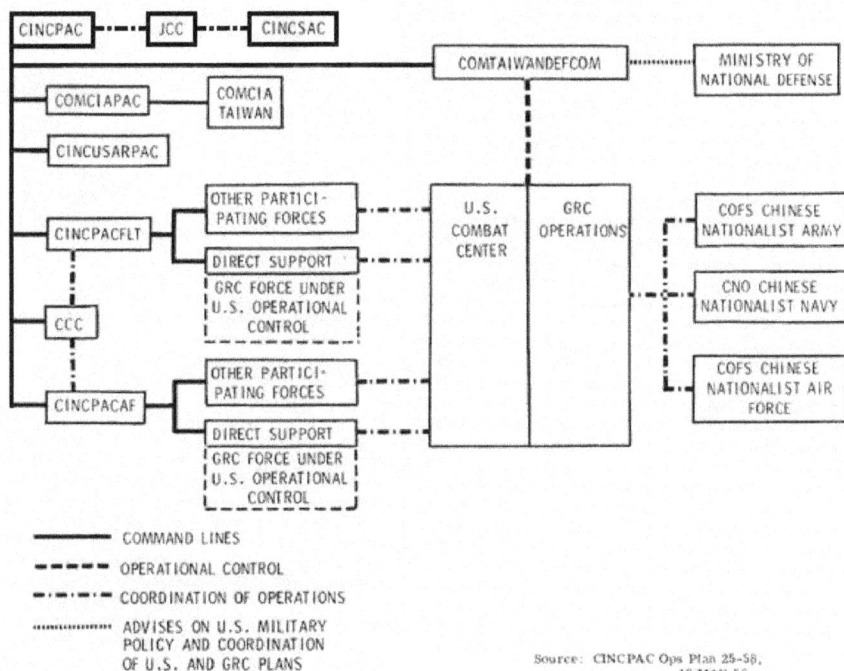

Source: CINCPAC Ops Plan 25-58, 16 MAY 58.

Admiral Stump had major responsibility for overall operational planning. On 16 May 1958 he published Ops Plan 25-58 to guide his major subordinate commanders in updating their respective operational plans. These subordinates were Gen. Laurence S. Kuter, Commander-in-Chief Pacific Air Forces (CINCPACAF); Vice Adm. Herbert C. Hopwood, Commander-in-Chef Pacific Fleet (CINCPACFLT), and Gen. Isaac D. White, Commander-in-Chief Army Pacific (CINCARPAC).[19]

Between 1955 and 1958 the United States made no basic changes in its Taiwan policies. This country would insure the military security of Taiwan and the Penghus (the defense of the offshore islands a matter of Presidential discretion), and it would support the Nationalist government as a free alternative to the Chinese Communist government. Maintaining Nationalist morale was deemed essential. However, frustrations on Taiwan mounted as Nationalists saw their hopes of recapturing mainland China fade, their position in the international community decline, and their national existence continue to be dependent on the United States. The determination of the Nationalists to hold the Kinmen and Matsu islands, now defended by more than 100,000 of their 450,000 regular ground troops, troubled U.S. policy planners. On the eve of a new Taiwan crisis, the nature of U.S. reaction to a Chinese Communist attack or an air and sea blockade of the offshore islands was still under discussion by the National Security council.[20]

II .. THE RENEWED COMMUNIST THREAT TO TAIWAN

There were signs of renewed trouble in the Taiwan Strait in July 1958 following the overthrow of the Iraqi government and the sudden dispatch of U.S. and British troops to Lebanon and Jordan, respectively to stabilize the situation in those countries.[1] For further details of these operations see Part Two of this monograph. In China, the Communists restricted western diplomats to the city of Peiping and intensified the propaganda war, especially their threats to "liberate" Taiwan. After four days of secret talks, Premier Khrushchev and Mao Tse-tung added to the tension by their joint communiqué from Peiping on 3 August demanding the withdrawal of U.S. and British forces from the Middle East.[2]

More ominous were signs of Chinese Communist military preparations. Heretofore unoccupied airfields in Fukien Province, across the strait from Taiwan, soon held an estimated 200 fighters, including MiG-17s. Air defense against CNAF reconnaissance and fighter sorties over the mainland became more effective. Between 23 June and 8 August, ten CNAF aircraft were shot down, seven in air engagements and three by AAA fire. Communist overflights of the Kinmen and Matsu Islands began. Across from the Kinmens the Communists augmented their artillery strength; on 18 August they fired about 100 shells at the islands. Three days later the war of nerves was intensified by the Soviet Union's Public pledge to assist the Peiping government's efforts to free Taiwan.[3]

Initial U.S. and Chinese Nationalist Reactions

Reacting to the Communist military buildup, the Nationalist government early in August declared a state of emergency for the offshore islands.4 It asked the United States to speed up military assistance, particularly F-86F aircraft and Sidewinder missiles, and reconsider a previously denied request for F-100D aircraft. The Nationalists also urged the deployment of USAF units to Taiwan.5 President Chiang Kai-shek, in a meeting with American Ambassador Everett F. Drumwright and the new commander-in-chief Pacific, Adm. Harry D. Felt, stressed the gravity of the crisis, decried U.S. and British failure to act decisively in the Middle East, especially in Iraq, and warned that this would encourage an attack on Taiwan. Reminded of his treaty commitments with respect to military action against the Communists (i.e., the 10 December 1954 notes), Chiang resentfully promised that his government would honor them. [6]

By 1958, the CNAF operated almost five hundred jet aircraft including F-84s
Source: U.S. Air Force

60 F-86Fs were the first reinforcements.
Source: U.S. Air Force

Improvement of the Chinese Nationalist Air Force, which now had 87,000 men and 825 aircraft (497 of them jets), had previously been under consideration by U.S. authorities. On 7 August the Department of Defense approved sending 60 rehabilitated F-86Fs to the CNAF, and on the 15th the JCS decided that the CNAF "should be built up and maintained in such a condition that it is qualitatively superior to the Chinese Communist Air Force." Pending approval of this decision by the Secretary of Defense, commanders were directed to take all practicable steps to provide the CNAF with suitable aircraft and necessary pilot training. A week later the Secretary of Defense approved the diversion to Taiwan of six F-100D aircraft earmarked for NATO.[7]

On 15 July, the day U.S. troops began debarking in Lebanon, the Air Force had placed all of its commands on alert. Ten days later, when the Middle East crisis appeared to have subsided, Headquarters USAF authorized field commanders to decrease their alert status and resume crew and combat-readiness training. PACAF intelligence, meanwhile, concluded that the next most likely trouble spot would be in the Taiwan area, a prediction given substance by the movement of MiG aircraft to airfields opposite Taiwan and the growing belligerency of the Chinese Communists. On 6 August the Air Force directed commanders concerned to reexamine their plans to support CINCPAC's Ops Plan 25-58. Gen. Otto P. Weyland, TAC commander, was asked to review his command's capability to deploy CASF X-Ray Tango to the Pacific, while Lt. Gen. William F. Tunner of MATS and General Kuter of PACAF were asked to determine their capabilities to support this deployment. Weyland responded by transferring planning and operational responsibility for deployment of CASF X-Ray Tango to the Twelfth Air Force because the Nineteenth was preoccupied in the Middle East with its CASF Bravo force.[8]

Kuter distributed the main portions of PACAF's interim Ops Plan 25-58 to his subordinate commanders on 7 August. It provided for U.S. military assistance to the Chinese Nationalist forces in three phases: Phase I, patrol and reconnaissance activities (already under way); Phase I[, the defeat of the attacking force; Phase III, air operations to destroy the Chinese Communist capability to make war. The last phase would be conducted primarily by the Strategic Air Command (SAC) with the support of CINCPAC forces. Most important to Kuter was Phase II. Assuming conditions short of a general war and the use of atomic weapons by both sides, PACAF would be responsible for striking 32 of the 52 pre-planned targets. Initial atomic strikes would be launched from Dark AB in the Philippines and Kadena AB in Okinawa, with additional strikes as the situation dictated. The Thirteenth Air Force at Dark AB would handle detailed planning of air operations for both PACAF and Chinese Nationalist air units. The Fifth Air Force, with headquarters in Japan, would provide tanker, tactical, and reconnaissance support to the Thirteenth during the early stages of operations.[9] Phase II was amended on 18 August when the Air Force alerted five Guam-based B-47s that had Mark 5 weapons with C cores to insure the most effective instantaneous retaliatory strikes against coastal airfields. This move gave SAC a role in limited

The Nuclear Dimension. B-47s based in Guam had atomic weapons
Source: U.S. Air Force

war. It was uncertain, however, whether these plans could be fully implemented because the American ambassador to Japan, Douglas MacArthur II, warned that the Japanese government might ask the United States not to use Japanese bases in a war against the Chinese Communists.[10]

PACAF's Ops Plan 25-58 was completed with little time to spare. Kuter distributed Annex F, the logistical part, on 14 August. It required air units moving to forward bases to deploy only the minimum equipment essential to the success of their missions and to make maximum use of prestocked equipment.11 The plan itself was not published until 20 August, only three days before the Taiwan crisis broke into full force.

Bombardment of the Kinmens

On 8 August the Chinese Communists, using an estimated 300 guns unleashed a tremendous artillery bombardment against the Kinmen islands, firing more than 40,000 rounds on the first day, Communist planes strafed the islands and sank a Nationalist LST. Artillery fire on the second day was equally severe; for each of the succeeding five days it averaged about 10,000 rounds, sufficient to impose a tight artillery blockade, Stepping up its propaganda war, Peiping radio urged the Nationalist garrisons to surrender. Leaflets dropped on Little Kinmen boasted that the defenders were isolated and that the People's Republic of China had an atomic stockpile.[12]

When the bombardment began, Secretary Dulles warned the Peiping government not to seize the Kinmens or Matsu's lest such an attempt threaten the peace. To back up his warning, the JCS on 24 and 25 August ordered the carriers Essex in the Mediterranean and Midway at Pearl Harbor and their destroyer escorts plus other ships to speed to the Taiwan Strait to reinforce the Seventh Fleet. The Army was ordered to expedite the shipment of a Nike battalion to Taiwan and more modern equipment for seven Nationalist army divisions on the offshore islands. Air commanders were directed to be ready to assist Nationalist forces by striking coastal air bases if a major attack threatened the principal offshore islands, defined as being Big Kinmen, Little Kinmen, and five islands of the Matsu group. They were cautioned that probably only conventional weapons would be initially authorized, but they were also instructed to be prepared to make atomic strikes deep into Chinese Communist territory.[13]

U.S. intelligence concluded initially that the Communists were testing U.S. and Chinese Nationalist reaction and might try to seize one or more of the islands if not convinced that the United States would intervene. However, an amphibious force for possible invasion of Big or Little Kinmen had not yet been observed. Intelligence sources did not anticipate an immediate strike at Taiwan or the Penghus, for they did not believe that either the Peiping government or the Soviet Union wanted a major war. But they warned of the danger of a Nationalist attack against the mainland if pressure against the offshore islands became too great.14

Meanwhile, the United States prohibited retaliatory aerial strikes against the Communists. Chiang Kai-shek vigorously protested against this injunction as inhuman and unfair. Although he promised to consult with the United States before taking any military action, the Generalissimo doubted whether he could control the morale of his armed forces and the public for more than a few days unless he was granted authority to bomb Communist gun positions, airfields, and communication centers.[15]

When the bombardment continued into the second week, Secretary Dulles announced that the United States had not decided whether the artillery attack on the islands related to the defense of Taiwan and the Penghus pursuant to the Joint congressional resolution of January 1955.[16] This served to keep both Chiang and the Communists guessing as to whether the United States would assist the Nationalists in defense of their offshore islands.

Deployment of USAF Units

Shortly after the bombardment of the offshore islands began, USAF field commanders requested additional air strength in the Pacific, especially on Taiwan. On 23 August, the commander of ATF 13, Brig. Gen. Fred M. Dean, asked for the Fifth Air Force's 15th Fighter-Interceptor Squadron, stationed on Okinawa. Two days later, General Kuter urged the intermediate deployment of part of TAC's CASF X-Ray Tango force from the United States, specifically an advance command element, an F-100 squadron, a C-130 squadron, and part of a communication and control group. The CASF force for the Pacific was initially composed of two segments, CASF X-Ray Tango and CASF Yankee. Only CASF X-Ray Tango deployments were made during the 1958 Taiwan crisis. Six RF-101s already on Taiwan on rotation were included in the CASF.

These requests were not immediately approved. U.S. military forces were still in the Middle East, and the meaning of the Chinese Communist bombardment was still uncertain. Washington authorities did not authorize any major air deployments within or to the Pacific for several days until they had evaluated all intelligence data and the international consequences of possible U.S. involvement in the Taiwan area. [18]

As a result, about five days elapsed before the 16th Fighter-Interceptor squadron was authorized to move. On the 29th, all but two of its F-86Ds flew from Kadena AB, 0kinawa, to Tainan AB, Taiwan, and were combat-ready seven and a half hours after the movement order. 0n the same day TAC began the CASF X-Ray Tango deployment. Preceded by tankers sent to staging bases en route, part of the force departed from George and Hamilton AFB's in California. Brig. Gen. Alvin P. Tacon and members of his advance command element were the first to reach their destination, Dark AB, in the Philippines, on 1 September.[19]

Since the JCS had indicated that only non-nuclear weapons probably would be initially authorized to counter a Chinese Communist attack on the offshore islands, Kuter asked on 31 August for additional CASF units. Noting that one B-36 could carry more 1,000 pound HE bombs than an F-100 squadron, he also recommended alerting one SAC B-36 squadron. The JCS disapproved a portion of Kuter's request and the alert of a B-36 unit, but on 2 September it authorized the deployment of the remaining force known as CASF X-Ray Tango (Balance). Finally, Kuter received permission on 6 September to deploy one F-101A squadron to Okinawa and 12 Air Defense Command (ADC) F-104s to Taiwan. Although neither had been previously identified as CASF X-Ray units, the F-101 squadron was designated CASF X-Ray Tango (Augmented) because it belonged to TAC.[20]

The entire CASF deployment (for a list of CASF X-Ray Tango units and their home stations see App 1) beginning on 29 August and ending on 12 September, took place in three distinct stages, following each other so closely in time that they blended into one continuous operation. Because of the great distances, one or two air refuelings were required for F-100, F-101, and RF-101 aircraft flying from California to Hawaii, then to Midway or Gram en route to Okinawa or the Philippines. The Twelfth Air Force controlled overall deployment from an air operations center at its headquarters in Waco, Texas.[21]

The movement of CASF X-ray Tango from the United States to Taiwan, Philippines, Okinawa, and Japan was generally successful. Travel times varied from about two to six days, depending on a variety of circumstances. The tropical storm Grace which swept through Guam on 2 September delayed initial F-100 and C-130 flights for 24 hours. Eight aircraft aborted or were delayed en route for maintenance, some not arriving until 25 September.[22]

This large-scale deployment was not achieved without overcoming obstacles. B-57 and C-130 flights were hazardous because there was only one permanent weather-reporting station for the 2,200 miles between the California bases and Hickam AFB, Hawaii. Without accurate upper-air data, aircraft had to buck high-altitude winds and arrived with only minimum fuel reserves. Some arrived and departed without proper cargoes, in part because of insufficient information concerning the cargo-personnel load and destination of each plane. Seaweed equipment prestocked at en route bases was either inadequate or in poor condition. Time was lost during deployment because refueling tankers had to be sent first to forward bases. Finally, limited cryptographic and other communication facilities at these bases hampered security, coordination, and control.[23]

While the CASF X-Ray Tango fighter and reconnaissance aircraft flew all the way with the aid of refuelings, the ADC F-104s had to be disassembled and transported in C-124 transports. Under the code name Jonah Able the first F-104 Starfighter left Hamilton AFB, Calif., on 8 September, was reassembled at Taoyuan AB, Taiwan, and made its first flight on 12 September. All 12 aircraft were not in place and operational, however, until 19 September. Their mission was to fly combat air patrol.[24]

Lockheed F-104A-20-LO of the 83rd Fighter Interceptor Squadron at Taeyan Air Base, Taiwan, on September 15, 1958. Source: U.S. Air Force

ADC had not anticipated this deployment, which was primarily for psychological purposes. Although the Starfighters had departed in great secrecy, their arrival on Taiwan quickly became a matter of public knowledge. Kuter believed that this type of publicity would firmly convince both the Chinese Nationalists and the Communists that the United States would support the former with the latest weapons.[25]

Support airlift for CASF X-Ray Tango was provided by MATS Ops Plan 115-58 Double Trouble. Like TAC, MATS received orders on 29 August to begin the CASF deployment. Its original plan called for support of the CASF in a single "package" of 36 airlift trips. But the deployment in three stages rather than one and the inclusion of an additional F-101 unit made this plan inapplicable. MATS C-118s, C-121s, and C-124s made 81 trips (19 for the first stage, 52 for the second, and 10 for the third). In all, 1,472 passengers and 860.1 tons of cargo were carried. One C-124 was lost, crashing near Guam and killing its 6-man crew and 12 passengers. Although numerous changes in plans considerably increased transport needs, MATS was able to provide TAC with the necessary support.[26]

To airlift the F-104s, including personnel and essential cargo, it was necessary to make 20 C-124 and 4 C-97 trips. Only the wings, tail, and nose assembly of the

Starfighter had to be detached in order to fit it into the maw of the giant Globemaster. This was the first large-scale movement of high-performance aircraft by C-124s to a troubled area.[27]

The giant C-124 Globemaster II transports put an entirely new meaning on to the term 'airlift.' Source: U.S. Air Force.

After the deployment, PACAF had 68 more tactical aircraft available: 42 offensive fighters, or an increase of 24 percent; 14 bombers or an increase of 28 percent; and 12 defense fighters, or an increase of 6 percent. For a list of USAF units available to PACAF after deployments see App 2.

Adding reconnaissance, transport, tanker, and other supporting units, the augmentation totaled 123 aircraft. The number of PACAF atomic aircraft reached 183. On Taiwan, the center of activity, there were about 4,400 USAF officers and airmen on 29 September.[28]

Deployment of Navy, Marine, and Army Units

The major U.S. naval deployments to the Taiwan area consisted of the aircraft carriers *Midway* and *Essex* and their destroyer escorts. The *Midway*, under JCS directions of 25 August, left Pearl Harbor on the 27th, arrived northeast of Taiwan, and commenced operations on 4 September, about 10 days later. The *Essex*, ordered to the Pacific on the 24th, passed through the Suez Canal on 29 August ("escorted" by Soviet submarines) and joined the Seventh Fleet on 16 September, about 22 days after the movement directive was issued. The arrival of the two carriers, their destroyer escorts, plus other reinforcements, made this fleet the largest naval force assembled since the Korean War. The Seventh Fleet now had 6 carriers, 3 heavy cruisers, about 40 destroyers, a submarine division, and about 20 other supporting craft. The carriers, in addition to the *Midway* and *Essex*, were the *Hancock, Shangri-la, Lexington,* and *Princeton.* Their aircraft included Skywarriors, Skyhawks, supersonic Crusaders, and Tigers; about 96 could deliver atomic weapons. The heavy cruisers were the *Helena, Los Angeles* and *Columbus.*

(Defense Lion notes that all three heavy cruisers were equipped to fire Regulus I nuclear-tipped cruise missiles. At this time these cruisers were considered to be strategic assets.)

During the augmentation period the Seventh Fleet changed its position in the Taiwan Strait to cover general war targets farther north, requiring PACAF units to cover as many former Navy targets as possible. The Navy maintained peak strength in the Strait only briefly, for some elements, including carriers, were soon withdrawn in accordance with the fleet rotation plan.[29]

USS Midway (CV-41) heading for Taiwan in August 1958. On her flight deck are the large, twin-engined A3D Skywarriors bombers with six FJ-4 Fury fighter-bombers and six F3H Demon missile-armed interceptors. A group of piston-engined AD Skyraiders can be seen on the after section of the flight deck. Beside them on the right are a group of F-8U Crusader fighters. Source: U.S. Navy

The Marines deployed 56 aircraft of Marine Air Group 11 (MAG-11) from Atsugi NAS, Japan, to Taiwan. The Seventh Fleet commander, Vice Adm. Wallace M.

Beakley, recommended this move after the JCS, on 25 August, instructed Admiral Felt and the TDC commander, Vice Adm. Ronald N. Smoot, to reinforce the air defense of Taiwan. MAG-11 had been alerted on 24 August but most of the unit did not move to Taiwan for nearly two weeks awaiting the selection of a suitable airfield. An advance Marine party spent several days surveying airfields and negotiating with USAF representatives over the use of one to meet Seventh Fleet requirements. Naval and Marine commanders rejected the first two that were suggested because they did not have an 8,000 foot runway and their selection would create logistical problems. The Navy also insisted that MAG-11 be located in an area where it could not only help defend Taiwan but also provide air cover and support for the fleet and augment its carrier strike force. On 29 August, agreement was finally reached on a third airfield, Pingtung North, which could meet requirements.[30] With the arrival of the Marine aircraft on Taiwan, 10 FJ-4, 30 F4D, 2 TV-2, and 2 R4D aircraft were stationed at Pingtung North and 12 FJ4 temporarily at Tainan.

MAG-II deployment began partly to support a previously scheduled Marine training exercise, Land Ho. Twelve aircraft of the initial MAG-11 squadron left Atsugi on 31 August and arrived that day at Tainan AB to participate in Land Ho, but six, scheduled for Pingtung, were delayed for nearly ten days because of mechanical difficulties or typhoon Grace. The other two squadrons of the group flew from Atsugi to Pingtung nonstop on 6 and 7 September, respectively.[31]

The North American FJ-4 Fury.
Source: U.S. Navy

The support airlift between 29 August and 11 September carried about 500 men and 205 tons of cargo in 54 trips - 43 by Marine transports, 6 by Navy, and 5 by Air Force. A sealift from Yokohama brought an additional 1,150 men and 3,900 tons of cargo. The gap created in Japan's air defenses by the departure of MAG-11 was quickly filled by the transfer of MAC-13 from Hawaii to Atsugi NAS.32 MAG-11 was ready for limited air defense operations on 1 September and on the 11th it began flying night cover sorties for the Nationalists and the U.S. Navy, who were resupplying the offshore islands. The 12 aircraft at Tainan rejoined the main Marine unit at Pingtung North, and MAG-11 became fully operational about 18 September. Taking into account all the obstacles encountered, the Marines considered most noteworthy the deployment of MAG-11 and its achievement of a limited operational status 10 days after the selection of an airfield.[33]

The JCS instructions of 25 August called for the expeditious shipment of a Nike-Hercules battalion to Taiwan. Certain types of modern equipment for the seven Nationalist divisions stationed on the offshore islands were also required within 60 to 90 days. As operational plans had not provided for the deployment of a Nike-Hercules unit, a month passed before it left Fort Bliss, Texas. Meanwhile, a battalion headquarters and four firing battery sites were selected and construction of semi permanent facilities began at Taoyuan on 15 September. The advance party of the 71st Artillery, 2d Missile Battalion, arrived by air on 9 September, and the remainder of the battalion arrived by ship on 9 October. It did not become operational until 25 0ctober.[34]

Nike-Hercules missiles deployed. For their time, these were powerful and capable weapons that were capable of carrying a nuclear warhead. They also were an accurate means of delivering that nuclear warhead to a land target.
Source: U.S. Army

III.   U. S. MILITARY PREPARATIONS ON TAIWAN

The heavy influx of aircraft, personnel, supplies, and equipment into Pacific bases during the first half of September created some initial confusion. Some air units did not arrive at their assigned destinations, upsetting logistical arrangements. Since more USAF aircraft went to Taiwan than planned, such as the ADC F-104s, air bases were overtaxed, requiring expensive crash improvement projects. Under TACs operational plan for CASF X-Ray Tango, the command element under General Tacon would have exercised operational, administrative, and logistical control. Because the CASF units had to be widely dispersed, General Kuter scrapped this plan, split up the command element, and integrated its personnel into the Thirteenth Air Force. Only a small CASF headquarters, redesignated the Thirteenth Air Force Command Element, Ryukus, remained at Kadena AB, Okinawa.[1]

Planning for Non-nuclear Operations

Some of these adjustments reflected in part the lack of firm operational plans at the beginning of the Taiwan crisis. There was insufficient time after the PACAF-TAC agreement of May 1958 to exchange information. The delay in obtaining JCS approval of PACAF's Ops Plan 25-58 prevented Kuter from completing and distributing this plan until shortly before the Communists began their bombardment on 23 August, and subordinate commanders had little or no time for detailed planning. More important, however, was the belated formulation of U.S. policy on the defense of Taiwan, the Penghus, and the offshore islands. Operational planners were not prepared for the JCS instruction of 25 August that required them to assume more responsibility for the air defense of Taiwan and the Penghus than they had anticipated and also to assist the Nationalists in defending the principal offshore islands, initially without nuclear weapons. This led to changes in plans, modification of the command structure, and a new disposition of air units.[2]

On the same day -- 25 August -- Admiral Felt informed his major subordinate commanders that he intended to prepare a special non-nuclear Annex H to his Ops Plan 25-58, and he asked them to take similar action. Observing that his original plan envisaged the employment of nuclear weapons and that the accompanying logistical system would be inadequate for non-nuclear operations, he requested an immediate appraisal of the non-nuclear capabilities of his commands.[3]

The possible U.S. injunction against nuclear weapons caused Felt and his subordinates much anxiety. They believed that the Communists had sufficient aircraft to stop the resupply of the offshore islands, U.S. airpower would be necessary to prevent their capture, and a U.S. non-nuclear defense might not be successful. Kuter was particularly apprehensive about conducting non-nuclear operations that would require a high sortie rate and weaken his ability to support his emergency war plan. He believed insufficient effort had been made to convince the National Security Council that the most effective way to deal with the enemy's numerical superiority was to use nuclear weapons. However, if non-nuclear operations had to be undertaken, he recommended that U.S. participation be limited to striking Communist coastal airfields opposite Taiwan.[4]

The fears expressed by field commanders concerning a U.S. non-nuclear policy in the Taiwan area were to no avail. On 29 August the JCS defined three new intermediate phases of possible Chinese Communist action against the offshore islands, Taiwan, and the Penghus, and it forbade the use of atomic weapons during each phase. President Eisenhower made a more definitive statent of this policy on 5 September when he granted the JCS emergency authority to order the use of only conventional weapons against any major Communist assault on the off-shore islands.[5]

Meanwhile, Felt informed the JCS on 5 September that his non-nuclear forces consisted of four F-100 and one B-57 USAF squadrons, stationed at Dark, Kadena, Naha, and Chiayi Air Bases, and aircraft on four Navy carriers. The squadrons could fly 170 sorties and deliver 400,000 pounds of bombs daily, and each carrier could launch up to 64 sorties and deliver 265,000 pounds of bombs daily, but in bad weather each could manage only 12 sorties per day. The Air Force units had a 30 day supply of POL, spare parts, support items, non-nuclear or "iron bombs" and 20mm ammunition but only a 15 day supply of external fuel tanks and pylons for F-100s. Bomb supplies aboard Navy carriers were sufficient for only 24 to 48 hours but were backed by an 18-day reserve in the Pacific area. A substantial amount of aircraft ammunition was available and logistical support for the carriers was adequate. The CNAF could fly 650 bomb-carrying sorties daily. In his assessment of his non-nuclear strength, CINCPAC did not specify the nature or location of Communist targets or the expected aircraft attrition. CINCPAC's hastily-assembled report on USAF capabilities appeared unduly optimistic in the light of subsequent analyses of the supply reserves in the Pacific.[6]

On 11 September, Felt issued his non-nuclear Annex H. This document and similar annexes prepared by subordinate commanders sought to define more precisely possible Communist action and U.S. and Nationalist reaction in the Taiwan area in order to control, if possible, a threatening military situation without resort to nuclear weapons. CINCPAC's original Ops Plan 25-58 had included Phase I (patrol and reconnaissance) and Phase II (defeat of the attacking force). It had anticipated a possible "transition phase" between the two. Annex H spelled out three intermediate phases of conflict between phases I and II as defined by the JCS on 21 August.

In Phase I-H, when there was no indication that the communists would attempt to capture the principal offshore islands, U.S. forces would provide only logistical assistance to the Nationalists, In Phase II-H, when the Communists would attempt to capture one or more of the principal offshore islands, U.S. forces would assist the Nationalists in attacking communist invasion forces, artillery positions, and airfields in the vicinity of the islands under attack. In phase III-H, when the Communists would extend the battle to Taiwan and the Penghus or to international waters close to both areas, U.S. and Nationalist aircraft would attack Communist airfields, GCI sites, military control centers adjacent to coastal airfields, and fixed targets. They would attack the airfields on a carefully controlled basis in a gradually expanding arc until they had destroyed all bases in an 800-mile radius capable of supporting Russian-built IL-28 Beagle aircraft.7

To coordinate the three intermediate phases of action, Admiral Felt designated the TDC commander, Admiral Smoot, as his subordinate unified commander on 11 September. Felt had maintained that the command structure on Taiwan had to be changed from a highly centralized control of a limited mobile nuclear force to less centralized control of non-nuclear units. To Kuter, such a change seemed unnecessary since it might break up the integrity of PACAF units. He also noted that the command structure would again need revision if nuclear weapons were authorized at a later date. On 10 September, however, JCS authorized the establishment of a subordinate unified command on Taiwan.[8]

COMMAND RELATIONS ON TAIWAN AFTER ESTABLISHMENT OF
A SUBORDINATE UNIFIED COMMAND
11 SEPTEMBER 1958

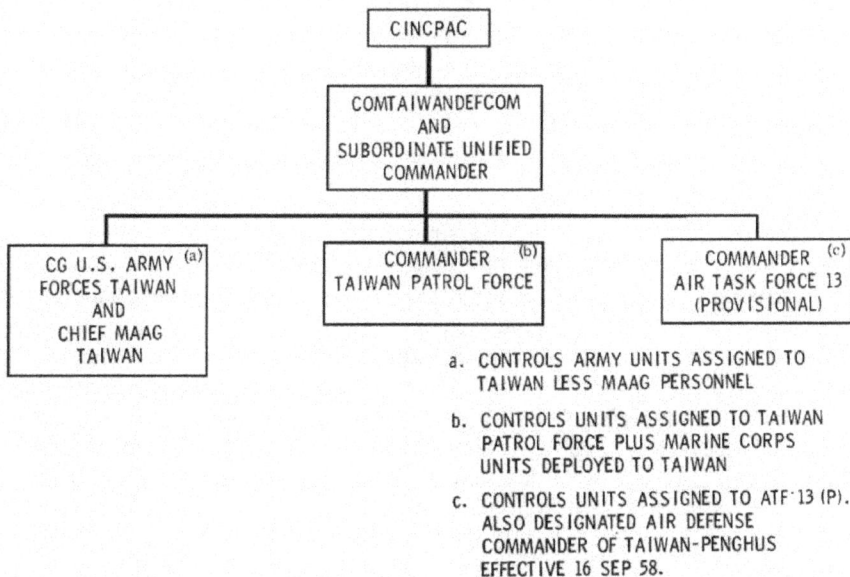

```
                        ┌─────────────┐
                        │  CINCPAC    │
                        └──────┬──────┘
                               │
                    ┌──────────┴──────────┐
                    │  COMTAIWANDEFCOM    │
                    │        AND          │
                    │ SUBORDINATE UNIFIED │
                    │     COMMANDER       │
                    └──────────┬──────────┘
         ┌─────────────────────┼─────────────────────┐
┌────────┴────────┐   ┌────────┴────────┐   ┌─────────┴────────┐
│ CG U.S. ARMY (a)│   │ COMMANDER   (b) │   │ COMMANDER    (c) │
│ FORCES TAIWAN   │   │TAIWAN PATROL    │   │ AIR TASK FORCE 13│
│      AND        │   │    FORCE        │   │  (PROVISIONAL)   │
│  CHIEF MAAG     │   └─────────────────┘   └──────────────────┘
│    TAIWAN       │
└─────────────────┘
```

a. CONTROLS ARMY UNITS ASSIGNED TO TAIWAN LESS MAAG PERSONNEL

b. CONTROLS UNITS ASSIGNED TO TAIWAN PATROL FORCE PLUS MARINE CORPS UNITS DEPLOYED TO TAIWAN

c. CONTROLS UNITS ASSIGNED TO ATF 13 (P). ALSO DESIGNATED AIR DEFENSE COMMANDER OF TAIWAN-PENGHUS EFFECTIVE 16 SEP 58.

Smoot would guide the Nationalists on current or projected military operations, establish and operate a combined operations center (COC) on Taiwan, and assume operational control of specified U.S. forces. He would also conduct patrol and sea and air reconnaissance in coordination with Nationalist forces. His three major subordinates on Taiwan were the commanding general of U.S. Army Forces Taiwan (who was also chief of MAAG Taiwan); the commander of the Taiwan Patrol Force (Navy; and the commander of ATF 13 (Air Force).

As the designated air defense commander under Smoot, the commander of ATF 13 -- General Dean-- would carry out air defense measures (including defense of naval units on request) and protect Nationalist ships resupplying the Kinmens and

Matsus. These operations would be conducted in accordance with limitations imposed by the U.S. policy of avoiding hostilities except as necessary to support the Nationalists and in self-defense. Operations conducted with any other forces of the Pacific Command in support of the Nationalists and by any Nationalist forces placed under U.S operational control would also be coordinated through the Taiwan C0C.[9]

Kuter and his Army and Navy equivalents relinquished to Smoot the operational-control of their units. They were responsible, however, for providing emergency logistical support to these units and to Nationalist forces and for furnishing facilities and personnel to the U.S. sector of the COC. Kuter's manifold duties also included augmenting the CNAF's ground environment system, providing personnel for ACW systems and other communication nets on the island, and bringing to operational readiness the coordination-centers in the Philippines and Japan.[10]

Strengthening Air Defenses

The Taiwan crisis did not expand beyond intermediate Phase I-H. Overt conflict was limited to Nationalist and Communist artillery exchanges, occasional aerial battles over the Taiwan Strait, and small sea engagements as the Nationalists tried to resupply the offshore islands. Free from direct participation in military action, the U.S. forces could spend the critical weeks of September and October in advising and assisting the Nationalists and in strengthening their own military posture in the Taiwan area.

On 16 September, Admiral Smoot assumed responsibility for the air defense of Taiwan and the Penghus and designated General Dean as his air defense commander. An exchange of Letters between Smoot and the chief of the general staff of the Ministry of National Defense, Gen. Wang Shu-Ming, confirmed previous U.S. and Nationalist verbal agreements on air defense arrangements. Most of the CNAF could be used to defend the offshore islands against attack, provide aerial cover for surface resupply efforts, or strike at Communist mainland targets, if necessary. About 121 aircraft were eventually assigned air defense missions: 49 USAF F-104s, F-86s, and F-100Ds; 56 Marine FJ-4s and F4Ds; and 16 CNAF F-86s and F-84Gs. These planes were stationed at Chiayi, Hsinchu, Tainan, Taoyuan, and Pingtung Air Bases on Taiwan.

Dean did not exercise complete control of the Marine aircraft, however. The commander of the Seventh Fleet, Admiral Beakley, insisted that they remain an integral part of his fleet because his carrier strike force, tailored for a nuclear conflict, was deficient in escort aircraft for non-nuclear strikes. After some controversy, MAG-11 was given a three-fold mission: air defense, which would have over-riding priority only if Taiwan and the Penghus were attacked or in imminent danger of attack; air support and cover for the Seventh Fleet; and augmentation of the fleet's strike force.[11]

The air defense of Taiwan called for 8 USAF, 2 Marine, and 8 CNAF aircraft to be placed on 5 and 15 minute day alerts and 8 USAF and 2 Marine aircraft on 5 and 15 minute night, all-weather alerts. Fortunately, there was sufficient time to

train those pilots who had not yet attained the desired standard of operational readiness.[12]

Communication and Logistic Problems

The concentration of airpower on Taiwan quickly drew attention to the most critical U.S, deficiency in the Pacific -- communications. U.S. air units, depending largely on the Chinese communication net, faced formidable problems of language, security, and reliability. Overdassified messages, nearly all dispatched as "operational immediate" created lengthy delays. Circuits from Taiwan to the Philippines, Okinawa, and Japan were woefully inadequate. The lack of sufficient on-line teletype communications alone could have made it virtually impossible to conduct combat operations.[13]

Because U.S. plans for major improvements had been rejected as too costly, crash communication projects had to be undertaken. During September and October about $4 million worth of Army equipment and about $2 million of USAF equipment were flown in. Between 25 August and 25 October, communication circuits on Taiwan increased from 25 to 200, including 120 voice and 30 teletype on-island circuits and 20 and 30 teletype off-island circuits. About 655 CASF, AACS, and other personnel arrived for temporary duty on Taiwan during the critical period to assist in manning communication facilities.[14]

The prospect of engaging initially in non-nuclear operations also required immediate adjustments in logistical requirements. During the crisis PACAF could make limited use of the Seaweed stocks earmarked for its emergency war plan, but Kuter was apprehensive about the adequacy of the stocks. Up to August 1958, stockpiling of Seaweed items was only 75 to 77 percent complete and of "absolute essential equipment" items only about 63 percent complete. Some very essential items were not stocked at all. The critically-short items included certain types of conventional ammunition. Because of these shortages and the prospect of a non-nuclear war, the Air Materiel Command's supply retrenchment program in the Pacific was halted. [15]

To meet the urgent need for supplies and equipment, the Fifth Air Force airlifted about 1,100 tons of essential items to Taiwan during the first two weeks of the bombardment. Thirteenth Air Force transports also brought vitally needed items. And large quantities of supplies were dispatched by sea. Starting and refueling units, war consumables, housekeeping supplies and equipment, and a tactical hospital comprised the bulk of these items. From PACAF and the United States came Sidewinder and other non-nuclear rockets, ammunition, and conventional bombs. External fuel tanks were also needed, and at least 3,500 tanks were flown from Korea and Japan to Okinawa and the Philippines, while some were airlifted from the United States.[16]

The stocked items were not only insufficient but some were in poor condition. Some were also at the wrong bases and had to be airlifted to where they were most needed. This required considerable intratheater airlift and time, and there was no assurance that both would always be available in a crisis. USAF officers on the scene believed that USAF units could not have sustained more than three days of

non-nuclear operations, and that the Navy's carrier strike force was in a similar predicament.[17]

While MATS was diverting much of its effort in early September to support the deployment of CASF X-Ray Tango and other units, a serious backlog of channel airlift items accumulated at Travis AFB, Calif., the main departure point. Channel airlift items are those supplied by a regular supply airlift as distinct from the airlift of the aircraft, supplies, and equipment and supplies of a tactical unit. To improve this situation, about the middle of the month, MATS began to use commercial airlift, letting contracts to 10 commercial lines for 807 tons but, not for an additional 179 tons because of excessive costs. Through commercial airlift and the gradual return of many MATS aircraft to the regular runs the backlog at Travis was reduced, but the problems persisted until the end of October. By then, 5,554 tons had been moved to the Pacific.[18]

The backlog at Travis reflected certain deficiencies in airlift planning and operations for the Taiwan crisis. All the services had failed to inform MATS fully of their greatly increased supply requirements. Units constantly upgraded their supply requisitions because of new priorities. MATS delayed procurement of commercial airlift and occasionally, items were brought by air that should have gone by sea.[19]

IV .. THE CONTROLLED CONFLICT

The "War" In The Air

The deployment of U.S. air units to the Pacific and augmentation of base facilities and supplies occurred against a background of limited air and sea dashes between Chinese Nationalist and Communist forces in the Taiwan area. Under the rules of engagement laid down by the United States for Phase I-H, the CNAF, in self-defense, could only attack Communist aircraft while they (CNAF) were on patrol and reconnaissance missions or defending the offshore islands and the convoys engaged in resupply operations.[1]

Whether the United States could retain control over CNAF's operations was uncertain, Nationalist frustrations because of military restrictions increased early in September when the United States and Communist China agreed to discuss the Taiwan crisis through their representatives at Warsaw, Poland. Aware that the Nationalists feared possible U.S. concessions, American authorities continued to anticipate some unilateral action such as CNAF bombing of Communist artillery positions. Both Chiang Kai-shek and the Nationalist Assembly in Taipei strongly advocated such action.[2]

Happily, the U.S, fears were not realized. It was conjectured that the lack of a firm U.S. commitment on the defense of the offshore islands kept the Nationalist government generally amenable to the restrictions on its air operations. This restraint was matched by the numerically superior Chinese Communist Air Force, which flew primarily defensive patrols. Thus the anticipated struggle for aerial supremacy over the Taiwan Strait did not materialize but was limited to occasional air dashes between the two Chinese air forces.

The results were surprising. During the period of crisis, the CNAF, in about 25 separate air engagements, destroyed 32 aircraft, probably destroyed 3 others, and damaged 10. It lost only two F-86Fs and 2 F-84Gs. Most of the encounters were between CNAF F-86s and Communist Mig-17s.[3]

The first major air battle occurred on 8 September when 12 F-86s fought 12 Mig-17s, destroying seven and possibly two others. In three separate engagements on 24 September, the CNAF destroyed ten and damaged three. This victory was highlighted by the destruction of four MiGs by Sidewinder missiles with which 20 CNAF F-86s had recently been equipped. A third significant air battle on 10 October resulted in the destruction of five MiGs, one of which was rammed by an F-86. The larger encounters occurred when escorted CNAF reconnaissance flights were dispatched to photograph mainland or coastal targets, and the smaller ones during routine air patrols.[4]

Analyses of the air-to-air dashes indicated that neither side employed unusual fighter tactics; their tactics were similar to those used during the Korean war. CNAF routine patrols normally flee at 35,000 to 37,000 feet whereas Communist patrols flew at 40,000 to 42,000 feet. When anticipating action the CNAF used four flights made up of four aircraft each and kept other patrols in the area on call if needed. MiGs generally outnumbered the F-86s by about 3 to 1, but the Communists usually dispatched only a single flight to engage.

MiG-17 Shot Down
Source: U.S. Air Force

Although possessing an altitude advantage, MiG pilots frequently demonstrated a lack of skill by permitting F-86s to get on their tails. Nor were Communist pilots very aggressive; CNAF pilots could shake pursuing MiGs by making six or seven turns at lower altitudes, The Communists were also reluctant to pull more than two and a half or three and a half Gs. They always tended to make left turns at higher altitudes. Their steeper turns (two or three Gs) were made at lower attitudes, likewise to the left. Their discipline and teamwork were poor, permitting CNAF pilots to attack stragglers. Communist pilots used an afterburner on one and possibly two occasions to escape. Their inferior showings are generally attributed to poor Soviet training and to a Communist decision to limit operations to their coastal areas.

CNAF pilots, by comparison, were well-trained, confident, flew excellent combat formations, were eager to make "kills" and pressed every advantage. Clearly, the Chinese Communist Air Force had been highly overrated.[5]

The CNAF reconnaissance force consisted of seven RF-86Fs, 25 RF-84Fs, and one RB-57A. At the insistence of the United States late in September, the CNAF curtailed the activities of this force to lessen the provocation arising from flights too near or over the Chinese Communist mainland.[6]

Meanwhile, during the period of crisis, the CNAF received or was in the process of receiving two additional RB-57s, six F-100Fs originally earmarked for NAT0, many F-86Fs that had been scheduled for Korea and Japan, and 16 C-119 transports. Many of these aircraft, such as the transports, were on loan. By the time most of the transfers were effected and the pilots trained, however, the Taiwan crisis had passed.[7]

The "War" At Sea.

Considerably less successful than CNAF operations was the Nationalist's initial naval effort to break the artillery blockade of the Kinmens. These islands, inhabited by 40,000 civilians, were defended by 86,000 troops and 56 medium artillery , 252 light artillery, and 160 AAA weapons. With the Nationalists outgunned about 4 to 1, the JCS on 29 August expedited shipments of 8-inch howitzers from the United States and Okinawa to the offshore islands. On the same day the JCS ordered elements of the Seventh Fleet to convoy Nationalist resupply ships to the offshore islands when necessary. These convoys would be limited, however, to Big and Little Kinmen and the five principal islands of the

Matsu group. The Matsu group were not under siege, but were inhabited by 10,000 civilians and defended by 23,000 troops and 8 medium artillery, 84 light artillery and 108 AAA weapons. The Navy was also advised to be ready to turn over to the Nationalists up to 8 LCMs and 28 LCVPs.

On the basis of these instructions, the United States and the Nationalists quickly prepared a convoy plan providing for U.S. air and surface escort to within three nautical miles of the Kinmens if Communist batteries could be avoided within that range. Admiral Felt indicated that the immediate objective was to lift the siege of the Kinmens and to increase the Nationalist's counter-artillery fire. To forestall resupply, the Peiping government announced its sovereignty over the territorial waters to a distance of 12 nautical miles, a claim promptly ignored by the U.S. State Department.[9]

Between 1 August and 15 September the Nationalists made only four organized attempts to resupply the Kinmens. The first attempt on 6 September, during a lull in the bombardment, was successful, but three others in the following week were not. Communist artillery fire (averaging about 6,675 rounds per day), hazardous sea conditions, poor equipment, lack of skills in handling supplies, incompetent logistical planning, and "utterly unbelievable" military command relationships combined to defeat their effort. CNAF C-46s made a few airdrops, but these provided only token supplies since about half the items initially dropped were either lost or severely damaged. By the third week in September USAF Headquarters estimated that the total Nationalist deliveries by sea and air amounted to no more than one day's requirement, then estimated at 700 tons.[10]

To U.S. authorities on the scene, it was apparent that the Nationalists were exerting less than an all-out effort to break the artillery blockade, In Washington, the Office of the Secretary of Defense, the JCS, and the State Department were deeply concerned over the Nationalists failure to resupply the Kinmens. Suggesting that the Nationalists were being deliberately inept in order to draw the United States into conflict with the Communists, the JCS insisted that they demonstrate an ability to resupply the islands or prove that their difficulties were insurmountable despite U.S, training, advice, and assistance.[11] Underlying this insistence was the belief that the Warsaw talks would provide no solution to the Taiwan crisis, that the United States faced another "Berlin blockade" with Peiping applying limited but sustained military pressure to exact politician and diplomatic gains, and that the Nationalist government would probably collapse if forced to withdraw from the offshore islands. Meanwhile, the Nationalists continued to threaten unilateral action to prevent the loss of the offshore islands by default, demilitarization, or neutralization.[12]

With JCS prodding and U.S. advice and assistance, the resupply picture began to improve. On 19 September, three Nationalist ships successfully completed their missions under heavy fire, the largest number to reach the islands in one day. shortly afterward, Admiral Felt reported that a U.S. Navy beachjumper unit and a beach master, plus organized beach parties for receiving supplies, had helped improve supply handling. Better under water demolition work opened up additional beach areas and permitted greater dispersion of arriving vessels. With

U.S. Marine air cover, Nationalist C-46 airdrops increased in frequency and in deliveries. As with the resupply success at sea, the CNAF's improved airdrop performance was due principally to U.S. advice and assistance. The picture further brightened when estimated supply requirements for the Kinmens were revised downwards to 300 tons and then 200 tons per day. In view of these developments, on 25 September, Admiral Felt instructed Admiral Smoot to undertake a massive resupply effort using all types of fishing boats, junks, and CNAF and Chinese Nationalist Civil Air Transport aircraft.[13]

Although Communist land-based artillery and gun and torpedo boats inflicted losses on the Nationalists, amounting to one LST torpedoed and sunk with two more damaged, three LCMs sunk and one swamped, two PCs damaged and three LVTs sunk with five more damaged, resupply during the last week of September proved highly successful. By month's end a total of 3,400 tons had been delivered to the beleaguered islands, 2,560 tons by surface ships, 210 tons by native junks, and 630 tons by air. The CNAF appeared able to airdrop more than the 200 tons required daily. When the first large-scale daylight airdrop drew Communist aircraft and an intense bombardment, U.S. and Nationalist officials agreed to nighttime aerial resupply, with daytime missions to be conducted only at irregular intervals.[14]

Early in October the chief of the U.S. Army Section, MAAG Taiwan, verified the fact that the Nationalists had exaggerated the critical nature of the supply situation. On a visit to Big Kinmen he found no grave supply problem; surprisingly little damage from 430,000 rounds of artillery except in the area of Nationalist batteries; low casualties; spectacular improvements in beach operations, with junks proving very successful; and morale, activity, and appearance of the Nationalist soldiers better than before 23 August.[15] This encouraging report and the advent of a temporary cease-fire in the Taiwan Strait presaged the end of the resupply problem.

The Communist Cease-Fire

Expectation that the Chinese Communists would resist the successful resupply efforts proved unfounded when the Communist Minister of Defense, Marshal Peng Teh-huai, suddenly ordered a one-week suspension of shelling "out of humanitarian considerations." Calling for a reconciliation with the Nationalists, he "authorized" them to resupply the offshore islands provided they did not use U.S. Navy escorts. The United States quickly agreed to halt its convoying operations but warned they would be resumed if the Communists reopened fire on the islands. The Nationalists also reluctantly agreed to stop firing but considered the suspension only the calm before a new storm.[16] During the artillery duels between the Chinese Communists and Nationalists from 23 August to 6 October the following rounds were fired: against all the Kinmens, 588,000; from the Kinmens, 88,045; against the Matsus, 183; from the Matsus, [13.]

In effect, the Communists were acknowledging that the artillery blockade of the offshore islands had been broken and that they were not prepared to reimpose it or to capture the islands. U.S. intelligence believed that the Communists, motivated by psychological and political considerations, were attempting to elicit U.S. proposals concerning the offshore islands in order to strain U.S. and Nationalist relations. They were also tacitly urging defection to the Communist cause and emphasizing the civil war nature of the conflict. On 13 October the cease-fire was extended for a week, permitting the Nationalists, who were strongly protesting U.S. suspension of convoy service, to continue unhampered resupply of their islands.[17]

240mm Howitzers Provide A Convincing Argument
Source: U.S. Army

The cease-fire gave U.S. authorities an opportunity to exert greater pressure on the Nationalist government to evacuate or at least reduce military garrisons on the offshore islands, particularly on smaller rocks like the two Tans (of the Kinmen group) regarded as indefensible, of little strategic value, and logistical burdens.[18] When Secretary Dulles conferred with Chiang Kai-shek on Taipei during 20-22 October, he emphasized the Nationalist's loss of Free World support because of the continuing strife and urged Chiang and his government to display less belligerency. Dulles secured Chiang's consent to reduce the military forces on the offshore islands in exchange for two battalions of 240mm howitzers that would improve considerably the counterfire capability of the Nationalists. Dulles envisaged a reduction of 15,000 to 20,000 Nationalist troops after Communist pressure subsided.[19] In a joint communiqué issued at the end of their conferences, Chiang also agreed to forego the use of force to restore his government on the mainland, [20] another significant concession in the light of his previous utterances.

The arrival of Secretary Dulles in Taipei had coincided with the end of the cease-fire, allegedly because a U.S. Navy ship had "violated" the "truce terms" a charge quickly denied. This confirmed Dulles's belief that the Communists were prepared to use any pretext to break the cease-fire. Communist antiaircraft fire against the Nationalist airdrops was only moderate, however.21 Five days later the Communists placed the bombardment of the Kinmens on an every-other-day schedule and "permitted" the Nationalists to replenish their offshore islands on even-numbered days. Taking full advantage of Communist concessions, the Nationalists delivered some 43,200 tons of supplies during 0ctober—40,300 by surface ships, 1,500 by junks, and 1,400 by air. Tota1 CNAF resupply losses because of ground fire were two C-46s destroyed and two damaged.[22]

Redeployment

With the artillery blockade broken and the danger of an enlargement of the conflict decreasing, U.S. commanders began to consider redeployment of their forces from the Pacific. The withdrawal began at the end of 0ctober when Adm. Arleigh Burke, Chief of Naval 0perations, authorized Admiral Felt to reduce his naval strength in the Taiwan Strait. Felt was instructed to leave two aircraft carriers and supporting ships on station and send the others to Subic Bay in the Philippines or to Yokosuka in Japan. Major redeployment of air units did not begin until December.

Both Navy and Air Force authorities in the Pacific recognized the political and psychological implications of this withdrawal. They were certain that the Chinese Communists would play it up as evidence of U.S. abandonment of the Chinese Nationalists and that the Nationalists would probably feel militarily weakened. They agreed, therefore, to couple any reduction of air strength with improvement of the CNAF, especially its air defense and all-weather capability, Kuter believed that Mig-19s, Mig-2ls, and a Soviet version of the Sidewinder would surely appear in the Chinese Communist air force inventory at an early date.[24]

The deployment of B-57C aircraft provided a potent night intruder capability
Source: U.S. Air Force.

The Air Force and Navy did not agree, however, on publicity for the return to the United States. Kuter strongly recommended a highly publicized movement of TAC B-57s, F-101s, and F-l00s through Japan, Korea, Taiwan, Thailand, and the Philippines to show that the United States was withdrawing with confidence and strength. He believed this would also rectify a previous imbalance in service publicity since the movement of Navy carriers to the Taiwan Strait at the beginning of the crisis had received much public coverage and the Air Force deployments relatively little. The USAF believed that the country tours should be restricted to Thailand and the Philippines. Although Kuter was supported in principle by Headquarters USAF, Felt recommended no country tour by units and as little publicity as possible. The issue was resolved early in December when the State Department supported Felt's position.[25]

In preparation for the redeployment, Felt transferred responsibility for the air defense of Taiwan and the Penghus, assumed by the United States in September, back to the Nationalist government.26 PACAF and TAC agreed late in November on a Twelfth Air Force redeployment plan (Ops Plan 37-58). CASF F-100 and F-101/RF-101 units would fly to Andersen AFB, Guam, where the Twelfth Air Force would assume operational control of the units. B-57s would go directly to Wake Island and from there to Hickam AFB, Hawaii. CASF tactical units, redeploying according to a prearranged order of precedence, would be supported by two C-130 troop carrier squadrons and a KB-50 air refueling squadron.[27]

The movement of CASF units to the United States began on 9 December. As in deployment, the F-100s and F/RF-101s passing through Pacific bases had one or two air refuelings. The C-130 transports carried military coordinating teams and the command element in addition to unit support equipment and personnel. MATS aircraft also provided airlift. Redeployment of the tactical units was completed on 18 December.[28]

The redeployment was successful but it pointed up a number of problems. there was not enough time to distribute the Twelfth Air Force's Ops Plan 37-58. No provision had been made for weather reconnaissance aircraft between Hickam AFB and air refueling areas. Military coordinating center units were undermanned for 24-hour duty. The receiving base for B-57s had to be changed from Moffett NAS, Calif., to George AFB, Calif., because of unfavorable

F-86Ds Continued To Defend Taiwanese Airspace
Source: U.S. Air Force

weather conditions and inadequate facilities at the former base. The Twelfth Air Force operations center at Waco, Texas, had trouble encrypting messages to PACAF, and many "launch" messages arrived either with little time to spare or too late for implementation. The problem was sufficiently acute to warrant waiving the encryption requirement in order to insure a reasonable efficiency in redeployment.[29]

Meanwhile, F-86Ds, F-104s, and MAG-11 units remaining on Taiwan were soon subject to rotation. An F-86D squadron from Okinawa and elements of another from Tainan AB replaced a redeployed F-100 unit at Chiayi AB.. A Starfighter squadron on Okinawa replaced the F-104 unit at Taoyuan AB. Both Felt and Kuter desired F-104s on Taiwan for psychological reasons, and they hoped that the CNAF would soon receive these fighters. MAG-11units began to return to

Japan on 1 February L959, and by mid-March only a small Marine squadron remained on Taiwan. [30]

The Chinese Nationalist Air Force was upgraded with F-100Ds and F-100Fs
Source: U.S. Air Force

The redeployment underlined anew the importance of modernizing the CNAF in accordance with JCS decisions. In addition to aircraft transferred or being transferred during the Taiwan crisis, further progress in 1958 and early 1959 included approval and funding under the Military Assistance Program (MAP) of 6 F-100Fs, 80 F-l00As, 4 RF-l00As, and 4 RF-101s. Several of the reconnaissance aircraft were delivered by the end of 1958. The Air Force also considered transferring under grant aid the 16 C-119 transports on loan to the CNAF. No immediate action was taken on recommendations by Felt and Kuter to further augment the CNAF with 10 RF-84s, one F-104 squadron, one F-86D squadron, or other MAP-financed aircraft. [31]

V .. AN APPRAISAL

The Taiwan crises of 1954-55 and 1958 were both preceded by numerous indications of Chinese Communist pressure against the offshore islands. The initial U.S. response in each instance reflected an ambiguous policy toward the islands. In January 1955, this country hastily decided to evacuate Nationalist military and civilian personnel from the Tachens only two days after making a decision to meet the Chinese Communist threat by a show of air strength from Taiwan. In mid-August 1958, despite considerable evidence of renewed Communist pressure, the U.S. Government still debated what its military response and its public policy should be in the event of a blockade or an attack against the islands. Not until 21 August, two days after the artillery bombardment of the Kinmens began, did the JCS direct U.S. forces in the Pacific to prepare to assist in the defense of only the principal offshore islands. Almost a week elapsed after the beginning of the Communist artillery bombardment before Washington authorized the deployment of USAF units within or to the Pacific, Meanwhile, the military initiative was left in the hands of the Communists.

U.S. policy was based on the recognition that the importance of the offshore islands was political and psychological (to the Chinese Nationalists) rather than strategic and that their defense was a matter of Presidential discretion. Whatever the merits of this policy, it made military planning difficult. USAF commanders, understandably, chafed at the slowness of the "decision making machinery" in Washington after there was ample evidence, in their opinion, of an impending military threat in the Taiwan Strait. They were uncertain whether they would be required to "fight or bluff." The JCS subsequently agreed that political guidance had been inadequate for developing defense plans for the offshore islands.[2]

In the area of diplomacy, the United States followed the policy of keeping the Japanese government fully informed during the crisis about this country's utilization of military bases in Japan. This assuaged Japanese feelings sufficiently to make unnecessary a request for "per mission" to use them.[3]

The 1958 Taiwan crisis underscored a need for more expeditious and more definitive operational planning. PACAF did not issue its operations plan until shortly before 23 August, thus precluding detailed implementation by subordinate commands. Planning between PACAF and TAC was incomplete on the eve of the crisis. PACAF believed that TAC's failure to provide sufficient detail (such as the type of equipment and the number of people required) did not enable the Fifth and Thirteenth Air Forces to plan for adequate support.[4]

The unexpected JCS instruction not to use nuclear weapons during the initial stage of conflict upset previous planning assumptions, for air units were operationally and logistically tailored primarily for nuclear warfare. USAF and Navy commanders viewed the injunction with apprehension for they were dubious of the success of non-nuclear operations against a Chinese Communist Air Force of unknown quality but formidable size. Air Force commanders were particularly concerned lest such operations affect their emergency war plans. They considered a non-nuclear requirement as a step backward in the art of warfare and a regression in national defense policy.[5]

If U.S. policy makers intended to authorize only a few non-nuclear strikes in order to warn the Chinese Communists against enlarging the conflict, it might be argued that this anxiety was overdrawn and that there had been no change in U.S. weapon policy.[6] However, the JCS directive of 29 August strongly suggests that relatively extensive non-nuclear operations were envisaged. In Phases II-H and III-H, U.S. airpower would make only iron bomb attacks against a widening range of coastal airfields, GCI sites, military control centers, and fixed targets plus certain other airfields up to a radius of 800 miles.[6] Based on established USAF war planning factors, PACAF concluded that it would require four squadrons of F-100s (or B-57s) or 20 squadrons of F-101s operating from Kadena or Dark Air Bases to neutralize one Communist airfield with bombings every fourth day to sustain neutralization. The aircraft requirement was based on the assumption that the Chinese Communists had more than 200 jet interceptors on coastal airfields and concentrations of heavy- and medium-type antiaircraft defenses. Unless the Chinese Communist air force and antiaircraft defenses proved woefully weak, it appears doubtful that available airpower in the Pacific could have engaged in such sustained operatlons.[7]

Whether adequate, and timely augmentation of aircraft, supplies, and equipment would have been forthcoming is questionable. General Kuter thought not. Certainly the field commanders believed that neither their military planning assumptions nor their prestocked assets were compatible with the concept of operations for the Taiwan Strait area directed by Washington authorities. They also agreed that the most important "lesson learned" during the 1958 Taiwan crisis was the need for a firm U.S. nuclear weapon policy.[8]

Early use not an option
Source: U.S. Air Force

U.S. authorities banned early use of nuclear weapons undoubtedly because they were determined to keep military activity localized in the Taiwan Strait while working for a cease-fire. They believed that a U.S.-Chinese communist conflict could be kept from expanding into an all-out war if U.S. forces limited themselves to using only conventional bombs. The goal of localizing the conflict was achieved, but since there was no test of the concept that the use of non-nuclear weapons would minimize the danger of all-out war, its validity can only be conjectured. That the United States belatedly modified its weapon policy for the Taiwan area appears borne out by the strong reaction of military commanders to the prohibition against the initial use of nuclear weapons. It is also indicated in Admiral Felt's query in mid-crisis about his non-nuclear capability. His decision to prepare a non-nuclear annex to his operations plan, his establishment of a subordinate unified command on Taiwan to conduct non-nuclear operations, and his subsequent decision to amend all contingency operations plans to provide for non-nuclear warfare. For the JCS the iron bomb policy reemphasized the need for an awareness by military planners of political objectives and by political authorities of the implications of their decisions on military planning.[9]

Whether the command structure on Taiwan should have been altered in preparation for possible iron bomb operations remains debatable. Felt believed the need for a subordinate unified command was confirmed by subsequent events, but Kuter did not.[10] Once the change was made, it functioned satisfactorily, and the appointment of the commander of AIF 13, General Dean, to head the air defense system on Taiwan created no important difficulties. The major objection to the subordinate unified command was that in the event of nuclear war there would be insufficient time to revert to a more centralized command.[11]

Unit movements to and within the Pacific theater were achieved without major difficulty. TACs Twelfth Air Force believed that the CASF concept of deploying substantial, mobile, tactical forces to a troubled area in a short time appeared confirmed, although Kuter thought the deployments were not quite as rapid as adverbised.[12] The Marine commander of MAG-11believed that his unit displayed much greater mobility in the transfer from Japan to Taiwan than did CASF X-Ray Tango.[13] This seems a highly questionable comparison in view of the greater problems inherent in flying tactical aircraft over vast distances of the Pacific.

The deployments did not follow original operational plans in every respect, largely because the Air Force had earlier diverted some tactical and support units to the Middle East following the outbreak of the Lebanon crisis in July. For example, TAC reassigned responsibility for deploying X-Ray Tango units from the Nineteenth Air Force to the Twelfth Air Force; X-Ray Tango units left the United States in three closely related stages rather than in one stage; and the movements of 12 F-104 Starfighters and MAG-11 to Taiwan were not provided for in initial planning. [14]

MATS airlift for both CASF and non-CASF units was adequate, but its channel airlift was not and had to be augmented by commercial carriers although excessive costs precluded MATS from contracting for as much commercial airlift as was needed. Even this did not eliminate the substantial backlog of supplies that existed at the MATS Pacific coast terminal, Travis AFB, during September and October. The backlog varied, but a contributing cause was the failure of commands to inform MATS in time of their increased requirements. The supply problems also confirmed the need for high-speed surface transportation to the Pacific outposts.[15]

The U.S. injunction against early use of nuclear weapons gave renewed importance to the possession of well-equipped, well-stocked, forward bases. Military assets at these bases were seriously inadequate, with iron bombs, ammunition, and auxiliary fuel tanks topping the list of critically deficient items, in addition to insufficient spare parts and equipment. Many logistical experts believed that preparations for iron bomb operations would require a partial reversal of the trend toward direct resupply from the United States.[16]

Even more serious was the communication problem in the Pacific, especially on Taiwan. Fully recognized during the 1954-55 Taiwan crisis, deficiencies had not been greatly remedied by 1958. the Hardtack high-altitude nuclear tests of August 1958, which resulted in widespread disruption of radio signals, also pointed up the inadequacy and the vulnerability of the military communication net in the Pacific.[17]

The Air Force conceded that in public relations, the U.S. Navy was more successful during the Taiwan crisis, primarily because command responsibility in the Pacific gave it a preferred position. News media representatives were cleared by Naval officers. Navy information offices were also generally better staffed. USAF information officers claimed that releases on units and personnel were either withheld or 'pared to the bone' by the Office of the Assistant Secretary of Defense (Public Affairs). The Air Force believed many of its informational problems could be alleviated by raising the professional standards of its career information officers.[18]

President Chiang Kai-shek
Source: Republic of China
Government

The 1958 Taiwan crisis provided a good test of the relations between the United States and the Chinese Nationalists. Politically, the Nationalists proved more amenable to U.S. advice and practiced greater restraint than expected. The willingness of President Chiang Kai-shek to agree to a reduction of his Army forces on the offshore islands (in exchange for more artillery from the United States) and his "renunciation of force" statement with Secretary Dulles in October 1958 were milestones in U.S. efforts to ameliorate the problem of the offshore islands.

In Nationalist military operations there were pluses and minuses. The CNAF was tactically much superior to the Communist Air Force, and its pilots were quick to learn how to fly later-model aircraft and use more advanced weapons such as the Sidewinder missile. Its limited airlift capacity also contributed to alleviating the artillery blockade of the Kinmens. The CNAF demonstrated how a force that had been reasonably modernized and well-trained under the U.S. Military Assistance Program could aid in the attainment of U.S. objectives in a localized conflict. The CNAF was also overgenerous in providing USAF forces on Taiwan with supplies and facilities. Deficiencies were most apparent in Nationalist counter-artillery fire and in logistical planning for and surface resupply of the offshore islands, The Nationalist navy, in contrast with the CNAF performed poorly at first and required much U.S. urging, training and guidance before demonstrating effectiveness. A need for more compatible Nationalist and U.S. military planning was also evident.[19]

US Navy Carriers In The Limelight
Source: U.S. Navy

But SAC Held The Strategic Line
Source: U.S. Air Force

In summary, the Taiwan crisis of 1958 remained a localized conflict and subsided without loss of Nationalist-held territory to the Chinese Communists. The reasons given for the success of U.S. policy were varied and inevitably reflected service and command viewpoints. In Kuter's opinion, the full-speed convergence of the "massive Seventh Fleet" toward the Taiwan Strait and the deployment of TAC's and ADC's Century fighters scored the greatest psychological impact on the Communists. The naval movement had the most immediate influence only because it was the most publicized.[20] The U.S. Navy believed that its attack carriers and combat Marines were principally responsible for keeping the conflicts in both the Taiwan Strait and in Lebanon from spreading. This view was strongly challenged by the USAF Chief of Staff, Gen. Thomas D. White, who gave primary credit to the Strategic Air Command.[21] Regardless of service claims, it seems likely that the U.S. tactical show of force, backed by strategic airpower, deterred the Chinese Communists from enlarging the conflict after the artillery blockade of the Kinmens was broken. Other factors, that undoubtedly contributed to U.S. success included intelligence assessments of Chinese Communist probing operations and the diplomacy used to prevent the Chinese Nationalists from taking unilateral military action.

The ultimate success of U.S. policy in 1958 did not obviate the fact that the United States was confronted with great risks in the Taiwan Strait area. These risks were intensified by the requirement that tactical forces be prepared to conduct non-nuclear operations for which they were ill-prepared from the standpoint of aircraft, supplies, equipment, and facilities. There was evidence that U.S. weapon policy was not firm and that a balance in tactical and logistical strength to conduct non-nuclear as well as nuclear operations remained to be achieved. Failure to achieve this balance could deprive the United States of a choice of weapons – non-nuclear or nuclear -- to deal with a future military crisis. In view of the possible, even likely, recurrence of trouble in the Taiwan Strait, the lessons of the 1958 crisis held great significance for U.S. political and military leaders.

NOTES TO PART ONE

CHAPTER 1

1. Nicola A. Ziadeh, Syria and Lebanon (London, 1957), pp 13-45, 236-61; Subcontractor's Monograph, HRAF-46 Patai-6, The Republic of Lebanon (New Haven, 1956), pp 1-33, 93-112.

2. Ziadeh, pp 27-190, 236-61; Patai-6, pp 1-22, 35-80, 362-85, 521-66.

3. Ziadeh, pp 108-25, 191- 212; Patai-6, pp 542-62.

4. Ziadeh, pp 153-72; Patai-6, pp 546-51.

5. DA CX 90, USARMA Beirut to DEPTAR Washington, 161500Z May 58.

6. DA CX 215, USARMA Beirut to DEPTAR Washington, 032308Z Jul 58.

7. State 3940, McClintock to Sec/State, 16 May 58, 8 pm.

8. *Ibid.*; State 4408 & 4525, McClintock to Sec/State, 2 & 7 Jun 58.

9. State 3772, McClintock to Sec/State, 11 May 58, 4 pm.

10. State 4408 & 4525, McClintock to Sec/State, 2 &,7 Jun 58; DA CX 177, USARMA Beirut to DEPTAR Washington, 200910Z Jun 58; DAF msg, USAIRA to CSUAF, C-6, 081029Z Jul 58; Dept of State, Foreign Policy Briefs Vol VIII, Supp l, 25 Jul 58.

11. Foreign Policy Briefs, 25 Jul 58; State 4492, McClintock to Sec/State, 6 Jun 58

12. Foreign Policy Briefs, 25 Jul 58.

13. *Ibid.*

14 State 4707 & 159, McClintock to Sec/State, l4 Jun & 6 Jul 58.

15. State 226, 228, & 374, McClintock to Sec/State, 9 Jul, 9 Jul, & 14 Jul; Foreign Policy Briefs, 25 Jul 58.

16. DA CX 90, USARMA Beirut to DEPTAR Washington, 1615OOZ May 58.

17. State 3775, McClintock to Sec/State, 11 May 58; DAF 9697, JCS to TAC, SAC, AMC, 16 May; DAF 9701, Hq USAF, to TAC, 17 May 58; memos for C/S USAF from Dir/Plans (sgd Betts), subjs: Current Actions Relative to the Situation in Lebanon, & Lebanon Situation, 19 & 22 May 58; memo for Adm Burke from Joint ME Planning Cmte, subj: Meeting at Dept of State, 1115 Monday, 19 May 58, dtd 19 May 58; draft msg, JCS to CINCEUR, CINCNELM, & CINCANT, 23 May 58.

18. State 367, McClintock to Sec/State, 14 Jul 58; DAF 1292, DAIN 132609, McClintock to Sec/State, 141702Z Jul 58.

19. DAF 1287, USARMA, USAIRA, *et al.* to CSUSAF *et al.*, 141048Z Jul 58; DAF 1292, DAIN 132609, McClintock to Sec/State, 141356Z Jul 58.

20. DAF 1313, CNO to SPECOMME & 6th Flt, 142249Z Jul 58; J-3 S 29/1, 19 Jan 59.

21. DAF 1326, Hq USAF to AMC, 15 Jul 58.

22. J-3 S 29/1, 19 Jan 59; J-3 Sitrep 1 (ME), 16 Jul 58.

CHAPTER II

1. J-3 S 29/1, 19 Jan 59, Encl C.

2. DAF 9712, SPECOMME to CNO, 162001Z May 58.

3. *ibid.*

4. DAS 1435, SPECOMME to CNO *et al.*, 161901Z Jul 58; DAF 1802, USAFE ADVON to subcomds, OPLWT 8RM-1850, 081800Z Aug 58; 19th AF, Blue Bat Consolidated Final Report, Tab F; Report by Adm James L. Holloway, Jr., Review and Evaluation of Operations in Lebanon (PROJECT RECAP), 7 Nov 58; J-3 S 29/1, 19 Jan 59, Encl C.

5. Lebanon Final Report for Task Force 62 and Marine Corps Troops Assigned SPECOMME, 18 May 1958-3 Oct 1958.

6. Command Diary, Hq Second Provisional Marine Force, Fleet Marine Force, Atlantic, 22 Aug 58; DAF 1333, CNO to 6th Fleet, 150335Z Jul 58.

7. Marine Corps rpt cited in n 5; msg, 6thFlt to CNO, 150335Z Jul 58; DA CX 248, USARMA & USAIRA to DEPTAR, 181600Z Jul 58.

8. Marine Corps rpt, pp 23-24.

9. Command Diary of Marine Force, 22 Aug 58.

10. State 401 & 402, McClintock to Sec/State, 151933Z & 151906Z Jul 58.

11. State 430, McClintock to Sec/State, 16 Jul 58, 7 pm; DAIN 133756, USAmb Beirut to Sec/State, 16 Jul 58, 7 pm.

12. DAF 1393, USAmb Beirut to Sec/State, 151836Z Jul 58.

13. DAF 1388, SPECOMME to CNO, 160644Z, Jul 58.

14. Msgs, McClintock to Sec/State: DAIN 133756, 16 Jul 58, 7pm;- State 423; 162025Z Jul 58; State 428, 161733Z Jul 58; State 430, 16 Jul 58, 7 pm.

15. DAF 1423, 6thFlt to SPECOMME, 170740Z Jul 58; DAF 1453, Hq USAF (AFCIN) to SAC, 181903Z Jul 58.

16. Marine Corps rpt cited in n 5.

17. Command. Diary of Marine Force, 22 Aug 58; J-3 Sitrep 2, 16 Jul 58; App 4 (G-4 evaluation) to Command Diary Narrative for Period 1-31 Jul 58, Hq Second Prov Marine Force, 10 Aug 58.

18. Marine Corps rpt.

19. J-3 S 29/1, 19 Jan 59; DAF 1541 & 1642, Hq USAF (AFCIN) to AIRU *et al.*, 232330Z & 282213Z Jul 58.

20. DAF 1573, SPECOMME to CNO (Sitrep 12), 242231Z Jul 58.

21. J-3 S 29/1, 19 Jan 59.

22. DAF 1649, SPECOMME to CNO, 27211Z Jul 58; DAF 1509 & 1543, DA SCO & DA SX 5556, USAREUR to SPECOMME, 221559Z & 240144Z Jul 58.

23. DAF 1482, Hq USAF (AFCIN) to ADC, 200128Z. Jul 58.

24. J-3 Sitrep 3 (ME), 17 Jul 58; DAF 1454, Hq USAF (AFCIN) to ALMAJCOM, 182004Z Jul 58.

25. DAF 1454 & 1471, Hq USAF. (AFCIN) to ALMAJCOM, 182004Z & 191000Z Jul 58.

CHAPTER III

1. DAF 1326, Hq USAF to AMC, 151500Z Jul 58.

2. DAF 1293, Hq USAF to TAC, 150248Z Jul 58; Hist, Dir/Plans, 1 Jul-31 Dec 58, pp 45-46.

3. DAF 1297, TAC to Hq USAF, 150517Z Jul 58.

4. DAF 1320, TAC to C/S USAF, 151700Z Jul 58.

5. DAF 1329, TAC to C/S USAF *et al.*, TAC COC 587778, 151846Z Jul 58.

6. 19th AF, Blue Bat Consolidated Final Report, Tab A, Sec III. See also DAF 1343, 1351, 1359, 1362, 1366, 1386, 1409, 1913.. & 1920, TAC to Hq USAF, 152310Z, 152359Z,, 160500Z, 1690323Z, 161000Z 160845Z, 170530Z, , 172315Z, & 182240Z Jul 58; DAF 1918, Hq USAF to ALMAJCOM, 181039Z Jul 58; DAF 1453, 1486 &, 1497 Hq USAF to SAC, 181903Z, 20205Z, & 212104Z Jul 58.

7. J-3 S 29/1, 19 Jan 59, Enct C.

8. USAFE, Support of the Lebanon Operation, pp 342-44.

9. Section based upon 354th Tactical Fighter Wing, Blue Bat Final Report, pp 15-24. See also DAF 1359 & 1366, TAC to C/S USAF, OPS SACT 4 & 5 1610500Z & 1601000Z Jul 58; DAF 1560 & 1582, AFCIN to AIRU et al. 242200Z, & 252024Z Jul 58; memo 2 for ASST Dep Dir/War Plans from Col

R.C. Franklin, Jr, Tac Div, subj: Congressional Force Posture Presentation, 20 Jan 59.

10. Section based upon 498th Bombardment Squadron, AU-I4 Final Report, 19 Nov 58, pp 2, 6-7, 11-12, 31, Incl 3. See also DAF 1359, 1386, & 1913, TAC to C/S USAF, OPS SACT 4, 6, & 11, 160500Z, 161845Z, & 172315Z Jul 58.

11. Section based upon 19th AF, Blue Bat Final Rpt, Tab A, Sec III, & Tab H, Incl I. See also DAF 1359, 1386, 1400, 1416, 1407, 1920, 1922, 1409, 1430 to C/S USAF, OPS SACT 4, 6, 7, & 9, STADEP 2, 4, & 5, OPS SACT 8 & 10, 160500Z, 161845Z, 162355Z, 171000Z, 170250Z, 182240Z, 200110Z 170530Z, & 171837Z Jul 58; DAF 1362, TAC to USCINCEUR, l60323Z, Jul 58.

12. 512th Fighter-Interceptor Squadron, Final Report on Operation Blue Bat; USAFE, Support of Leb Op, pp 349-51; DAF 1321, USAFE ADVON to C/S USAF, ODCP 8RM-1459, 15150Z Jul 58.

13. 19th AF, Blue Bat Final Rpt, Tab D; memo for Under SAF from Maj Gen J.E. Smart, subj: Resume of Airlift Operations Pertaining to the Middle East, 8 Aug 58. See also DAF 1416 & 1430, TAC to C/S USAF, OPS SACT 9 & 10, 171000Z & L171837Z Jul 58; DAF 1433, TAC to OSAF, TIS 58 7805, 171939Z Jul 58.

14. Section based upon 19th AF, Blue Bat Final Rpt, Tab F; USAFE, Supp of the Leb Op; draft Chap IV, Hist, MATS, 1 Jul-31 Dec 58, p 2. See also DAF 9833, Hq USAF to SAC, TAC, AMC, 23 May 58.

15. DAF 1312, USAFE ADVON to 322 AD, #ODCP 8RM-1450, 151600Z Jul 58.

16. USAFE, Supp of Leb Op, p 355.

17. DAF 1330, USAFE ADVON to EAME, AACS, et al., OOT 8RM-1462, 151600Z Jul 58; DAF 1335 & 1336, USAFE ADVON to 17th AF & 3d AF, 151744Z & 151510Z Jul 58.

18. DAF 1360, USAFE ADVON to 7100 Support Wing et al., ODCP 8RM-1465, 151640Z Jul 58.

19. DAF 1337, USAFE ADVON to USAREUR , 151835Z Jul 58.

20. USEUCOM, Chronology of Operation Blue Bat, 4 Nov 58; USAFE Supp of Leb Op, pp 355-56; Smart memo of 8 Aug as cited in n 13; J-3 S 29/1, 19 Jan 59; 322d, Air Div, "History of Operation Blue Bat."

21. DAF 1424, SPECOMME to USCINCEUR et al., 170741Z Jul 58.

22. DAF 1447, SPECOMME to JCS, 171916Z Jul 58; USAFE, Supp of Leb Op, p 357.

23. DAF 1477 & 1497, Hq USAF to SAC, 19 & 21 Jul 58; JCS 1887/527, 19 Feb 59; DAF 1921, USAFE ADVON to C/S USAF, ODCPS 8RM-1592, 191725Z Jul 58.

24. Smart memo, 8 Aug 58; DAF, 1921, USAFE ADVON to C/S USAF, 191725Z Jul 58; DAF 1923, AFOOP to ALMAJCOM, 200611Z Jul 58; DAF 1596. USAFE ADVON to CINCEUR, 260950Z Jul 58; USAFE, Supp of Leb Op, pp 360,i 368.

25. DAF 1161, USAFE ADVON to 10th Tactical Reconnaissance Wing *et al* 262055Z Jul 58; DAF 1596, USAFE ADVON to CINCEUR, 260950Z Jul 58

26. Smart memo, 8 Aug 58.

27. Ibid.; USAFE, Supp of Leb Op, p 349, draft Chap IV, Hist, MATS, 1 Jul- 31 Dec 58; ltr, K.E. Crane, Hq MATS to Air University Historical Division Liaison office, subj: Questions on MATS participation in Blue Bat Operation, 13 Aug 59.

CHAPTER IV

1. Map in 19th AF, Blue Bat Consolidated Final Report, Tab A; USAFE, Support of the Lebanon Operation, p 353.

2. 19th AF, Blue Bat Final Rpt, Tab A.

3. USCINCEUR Report, Blue Bat Critique, 2-3 Dec 58.

4. 354th Tactical Fighter Wing, Blue Bat Final Report.

5. 498th Bombardment Squadron, AU-14 Final Report, 19 Nov 58.

6. 512th Fighter-Interceptor Squadron, Final Report on Operation Blue Bat.

7. 19th AF, Blue Bat Final Rpt, Tabs A and H; J-3 S 29/1, 19 Jan 59, Encl C

8. 19th AF, Blue Bat Final- Rpt, Tab E.

9. Lt Gen Sir John B. Glubb, *A Soldier with the Arabs* (New York, 1957), *passim*; Anne Dearden, *Jordan* (London 1958), *passim*; Raphael Patai, *The Kingdom of Jordan* (Princeton, 1958), *passim*.

10. J-3 Situation Report 3, 17 Jul 58; DAF 1404, USAmb Amman to Sec/State, 162355Z Jul 58; DAF 1423, 6thFlt to SPECOMME, 170740Z Jul 58.

11. DAF 1454, Hq USAF to ALMAJCOM, 182004Z Jul 58.

12. DAF 1541, AFCIN to AIRU *et al.*, 232330Z Jul 58.

13. For the USAFE lift to Jordan, see memo for Under SAF from Maj Gen J.E. Smart, Asst VC/S, subj: Resume of Airlift Operations Pertaining to the Middle East, 8 Aug 58; USAFE, Support of the Lebanon Operation pp 362-64; draft Chap III, Hist, Eastern Transport AF, I Jul-3I Dec 58, p 14; DAF 1560, AFCIN to AIRU et al., 2422200Z Jul 58; DAF 1562, 1564, & 1662, JCS to SPECOMME 24 Jul, 24 Jul, & 29 Jul 58; DAF 1596 USAFE ADVON to CINCEUR, 260950Z Jul 58; DAF 1622, SPECOMME to USAFE ADVON, 260121Z Jul 58; DAF 1623,1624, & 1750, SPECOMME to CNO, 272O31Z, 280556Z Jul, & 062221Z Aug 58; DAF 1563, USAFE ADVON to

SPECOMME, OX 8RM-1720, 291030Z Jul 58; DAF 1795; State 1398, Dep/State (Dulles) to AmEmb London (for PM), 2 Aug 58; J-3 Sitrep 11, 24 Jul 58; DAF 1674, SPECOMME to MIDEAST, 301611 Jul 58; DAF 1718, 1820, & 1939, MIDEAST to MINDEF, 012020, 062115Z, & 102010Z Aug 58.

14. DAF 1402, AMAIRFOR to SPECOMME, 162145Z Jul 58; DAF 1430, TAC to C/S USAF, OPS SACT 10, 171837Z Jul 58; DAF 1527, SPECOMME to CNO, 211341Z Jul 58.

15. USAFE, Support of the Lebanon Operation, pp 326, 333-34, 352; DAF 1527, SPECOMME to CNO, 211341Z Jul 58; DA 945322, DEPTAR to USCINCEUR, 23 Jul 58.

16. Report by Adm James L. Holloway, Jr, Review and Evaluation of Operations in Lebanon (PROJECT RECAP), 7 Nov 58, in Plans Div Files OPS 30, Leb, Sec 4.

17. Report by Adm James L. Holloway Jnr Operation "Blue Bat," 15 Jul-25 Oct 58, dtd 5 Dec 58, in Plans Div Files OPS 30, Leb, Sec 4.

18. DAF 1600 & 1720, SPECOMME to CNO, 251611Z Jul & 022116Z Aug 58.

CHAPTER V

1. DAF 2028, SPECOMME to CNO, 231811Z Aug 58.

2. DAF 2449, SPECOMME to CNO, 101456Z Sep 58.

3. DAF 1821, SPECOMME to CNO, 082216Z, Aug 58.

4. State 617 & C49, McClintock to Sec/State, 22 & 24 Jul 58.

5. DAF 1573, 1722, &, 2449, SPECOMME to CNO, 242231Z Jul, 021631Z Aug, & 101456Z Sep 58; State 433, 496, 698, & 744, McClintock to Sec/State, 16 Jul, 18 Jul, 25 Jul, & 26 Jul 58.

6. State 473, 496, 530, 621, 715, & 819, Murphy to Sec/State, 17 Jul, 18 Jul, 19 Jul, 22 Jul, 25 Jul & 30 Jul 58; State 533, 534 617, 642, 698, 883, 898, McClintock to Sec/State, 19 Jul, 20 Jul, 22 Jul, 23 Jul, 25 Jul, 30 Jul, & 2 Aug 58.

7. State 943, Murphy to Sec/State, 5 Aug 58, 6 pm; State 1248, McClintock to Sec/State, 22 Aug 58; Navy 8992-8934, AmEmb Beirut to AmEmb Amman et al., 021956Z Sep 58.

8. DA msg, AMLANFOR to SPECOMME, 020950Z Sep, 130B815Z Sep, & 162030Z Sep 58; State G-56, McClintock to Sec/State, 5 Sep 58; DA CX 316, USARMA, USAIRA, *et al.*, Beirut to DEPTAR, CSUSAF, et al.,081218Z Sep 58.

9. DA msg, AMLANFOR to SPECOMME, 210620Z Sep 58, 230530Z Sep, & 270505Z Sep 58; DA CX 336, USARMA to DEPTAR, 291336Z Sep 58; State 1682 & 170 McClintock to Sec/State, 24 & 25 Sep 58.

10. DA msg, ALFJB X-9-10-1, AMLANFOR to SPECOMME, 090545Z Oct 58; State G-80, 1773, 1921, 1951, 1966, 1967, 1988, 2009, & 2121, McClintock to Sec/State, 8 Oct, 1Oct, 10 Oct, 14 Oct, 15 Oct, 15 Oct, 16 Oct, 20 Oct & 28 Oct 58.

11. DAF 1592, Hq USAF (AFOOP) to SAC *et al.* , 252208Z, Jul 58.

12. DAF 1744 & 1747, JCS 946103 & 946107, JCS to SPECOMME, both 5 Aug 58; State 993, McClintock to Sec/State, 7 Aug 58, 11 pm; Navy msg, SPECOMME to CNO, 170226Z Aug 58.

13. J-3 Sitrep 32, 13 Aug 58; State 10192 , McClintock to Sec/State, 13 Aug 58.

14. J-3 S 29/1, 19 Jan 59, Encl C.

15. *Ibid.*; DAF 2547,.JCS to SPECOMME, 072216Z Sep 58; DAF 2577, NELM Rr Ech to CNO, 181746Z Sep 58; DAF 2510, 2807, & 2859, SPECOMME to CNO, 141101Z Sep, 071116Z Oct, & 121811Z Oct 58; State 2059, McClintock to Sec/State, 23 Oct 58, 11 am; DA msg, LANFOR to SPECOMME *et al.*, 250712Z Oct 58.

16. DAF 2538, SPECOMME to CNO, 162106Z Sep 58.

17. 354th Tactical Fighter Wing, Blue Bat Final Report; 363d Composite Reconnaissance Squadron, Final Report on CASF BRAVO Deployment; draft Chap IV, Hist, MATS, 1 Jul-31 Dec 58, p 6.

18. 322d Air Division, History of "Operation Blue Bat," pp 31-34.

19. DAF 2885, JCS to SPECOMME *et al.*, 161815Z Oct 58; DAF 2888, Hq USAF to USAFE, 162153Z Oct 58; DAF 2875, USAFE ADVON to TAC, ODC 8RM-2409, 140910Z Oct 58; CINCSEPCOMME Report, Operation "Blue Bat," 5 Dec 58; DAF 2859, SPECOMME to CNO, 121811Z Oct 58; DAF 2443, TAC to CSUSAF, TOOT-PC 588133, 082142Z Oct 58; DAF 2483 TAC to USAFE ADVON, 092022Z Oct 58; ltr, CWO G.L. Smith, 354th TFW to 9th AF, attn: NCFDM subj: Rotation Requirements, 4 Nov 58; Hist, TAC, 1 Jul-31 Dec 58, Chap II, pp 136-39, & Chap VIII, pp 191-97.

CHAPTER VI

1. JCS 2295/3, 9 Feb 59.

2. l9th Air Force, Blue Bat Consolidated Final Report, Tab A; 354th Tactical Fighter Wing, Blue Bat Final Report; 463d Troop Carrier Wing Mission Report, 15-24 Jul 58.

3. Draft Chap IV, Hist, MATS, I Jul-31 Dec 58; memo for Asst DCS/O, MATS from Col J.C. Gunnison, 26 July 58.

4. 19th AF, Blue Bat Final Rpt, Tabs A, H, E, & F.

5. J-3 S 29/1, 19 Jan 59.

6. State 2158, 2112, 2059, 617, 649, & 657, McClintock to Sec/State, 29 Oct, 27 Oct, 23 Oct, 22 Jul, 22 Jul, 24 Jul & 24 Jul 1958; DAF 2449, SPECOMME to CNO, 101456Z Sep 58.

7. 19th AF, Blue Bat Final Rpt, Tab A.

8. Ibid.; Gen Thomas D. White, in discussion following speech before National Press Club, II Jan 60.

NOTES TO PART TWO

CHAPTER 1

1. Dept/State, *United States Relations with China* (Washington, 1949), pp iii - xviii *Department of State Bulletin* 16 Jan 50 - p 79.

2. *Dept/State Bull.* 3 Jul 50, p 5; Hearings Before the Senate Cmte on Armed Services and the Cmte on Foreign Relations, 82d Cong, 1st Sess (1951), *Military Situation in the Far East*, p 903.

3. *Dept/State Bul1*, 9 Feb 53, pp 207-11.

4. House Rpt, *Treaty Provisions Relating to the Use of U.S. Forces for Mutual Defense,* 27 Dec 56, 84th Cong, 2d Sess, pp 26-30; Staff Briefing, 15 Sep 54, App 53 in Hist, FEALF, I Jul-31 Dec 54, Vol II.

5. A. Doak Barnett, *Communist China and Asia*, published for the Council on Foreign Relations (New York 1960), pp 411- 12.

6. Staff Briefings on Formosa and Southeast Asia, 15 Sep thru 1 Dec 54, Apps 53-57 in Hist, FEALF, 1 Jul - 31 Dec 54, Vol II; Senate Rpt 13, *Senate Miscellaneous Reports on Public Bills,* 84th Cong, 1st Sess (1955).

7. FEC/UNC Comd Rpt, Oct 54, pp 37 - 41; Hist, FEAF, 1 Jan-30 Jun 55, Vol I, Pt II, pp 19-22.

8. Senate Rpt 13, 1955.

9. FEC/UNC Comd Rpt, Oct 54, pp 37-4; Hist, FEAF 1 Jan - 30 Jun 55, Vol I, Pt II, pp 20-22; Special Rpt, Fifth Air Force in Formosa, 25 Jan - 1 Jun 55, pp 24 - 25.

10. Fifth Air Force in Formosa, pp 24 – 25, 36.

11. *Ibid.*, pp 36 - 42; Hist, FEAF 1 Jan - 30 Jun 55, Vol I, Pt II, pp 20 - 24; *New York Times,* 25 & 27 Feb 55.

12. Fifth Air Force in Formosa, pp 36 – 42; Hist, FEAF, I Jan - 30 Jun 55, Vol I, Pt II, pp 20 – 24; Special Rpt, Fifth Air Force in Taiwan, 1 Jun - 31 Dec 55, pp 10 - 11.

13. Fifth Air Force in Formosa, pp 30 – 32, 45 - 46, 69 - 97.

14. *Ibid.,* pp 45 - 55, 105; Hist, FEAF, 1 Jan - 30 Jun 55, Vol I. Pt II pp 28 – 29; Fifth Air Force in Taiwan, pp 10-11, 30 – 32; Hist, ATF 13, 1 Jul - 31 Dec 58, p 20; file 381, China, 29 Nov 50, Sec 3, in Federal Records Center (FRC).

15. Special Rpt, Fifth Air Force in the Taiwan Straits Crisis of 1958, dtd 31 Dec 58, p 14; Hist, PACAF, 1 Jul - 31 Dec 57, Vol 1, pt I, p 11, and 1 Jan - 30 Jun 58, Vol I, pt I, p 11; OCB Rpt, U.S. Policy Toward Taiwan and the Government of the Republic of China, dtd 16 Apr 58, in D/plans, Ops 20-9, 1 Jan - 25 Aug 58; AIIR IR-717-58, subj: Air Order of Battle for the Chinese Nationalist Air Force (as of 31Aug 58), dtd 30 Sep 58, in AC/S, Intel; Hist Div, 13th AAF, Special Rpt, Taiwan Straits Operations, p 63; Briefings by

Brig Gen Alvin P. Tacon, Lessons Learned in CASF Deployments to the Far East, at Senior Operations Officers Conference, Maxwell AFB, 29 - 30 Jan 59, pp 9 - 12.

16. *Semiannual Report of the Secretary of Defense*, 1956, p 266, & 1957, pp 301-2; Hist PACAF, 1 Jul – 31 Dec 58, Vol I, Pt II, pp 76 - 77.

17. *Semiannual Rpt. of SOD*, 1957, p 296.

18. File Ops 6-33, 1Jan – 12 Sep 58, in D/Plans; Hist; ATF 13, 1 Jul - 31 Dec 58, p 17.

19. CINCPAC Ops Plan 25-58, 16 May 58, to D/Plans.

20. OCB Rpt, US Policy toward . . . China; CINCPAC Ops Plan 25-58, p D-6; memo, Exec Secy, NSC to NSC, subj: The situation in the Taiwan Straits Area, 13 Aug 58, in D/Plans, Ops 2O-9.

CHAPTER II

1. For a discussion of this operation, see study by USAF Historical Division Liaison Office: Air Operations in the Lebanon crisis - 1958. *[Lion Defense Publications – this study forms Part One of this book]*

2. File Ops 22, 1 Jan 59, in D/Plans; *New York Times*, 4 Aug 58.

3. Hist, PACAF, 1 Jul - 31 Dec 59, Vol I, Pt I, pp 158 - 61, & Vol II, Doc 1, Rpt, Taiwan Straits Crisis, by Col H.B. Hull, Dec 58; AIIR IR-468-58, subj: Scope and Policy of Reconnaissance and Photography Missions for F-84s, 11 Jul 58, & AIIR IR-575-58, subj: Recap of CAF Aircraft Destroyed by CHICOM MiGs and AAA fire, 11 Aug 58, in AC/S, Intel.

4. *New York Times*, 5 Aug 58.

5. Msg, TDC/MAAG to CINCPAC, 4 Aug 58; msg 1757, Hq USAF to ALMAJCOM, 5 Aug 58, in FRC.

6. Msg 127 Taipei to Secy/State, 4 Aug 58, in FRC.

7. CINCPAC Ops Plan 25-58, 16 May 58, p D-6; msg, Hq USAF to CINCPACAF, 7 Aug 58, in D/Plans, Ops 20-5, 5 Jun - 21 Aug 58; msg JCS 946950, COFSUSAF to CNO, COFSA, *et al.*, 22 Aug 58, in FRC; Hist, D/Ops, Hq USAF, 1 Jul - 31 Dec 58, pp 42 - 43.

8. Special Rpt, Fifth Air Force in the Taiwan Straits Crisis of 1958, p 1; msgs 1992 & 1757, Hq USAF to ALMAJCOM, 25 Jul & 6 Aug 58, in FRC; Presentation by Gen Laurence S. Kuter, at ZI Commanders Conf, Patrick AFB, 20 - 21 Nov 58; Hist, MATS, 1 Jul – 31 Dec 58, Chap IV.

9. Msg, CINCPACAF to ALMAJSUBCOM, 7 Aug 58, in D/plans, Ops 20-9; msg 1971, CINCFACAF to COSUSAF, 19 Aug 58, in FRC.

10. Msg 1961, Hq USAF to CINCPACAF (personal LeMay to Kuter), 18 Aug 58; msg 1971, CINCPACAF to COFSUSAF, 19 Aug 58, both in FRC; msg 361, Tokyo to SECY/STATE, 18 Aug 58, in D/Plans, Ops 2O-9.

11. Hist, 5th AF, 1 Jul - 31 Dec 58, p 1O8.

12. Hull Rpt as cited in n 3; Hist, ATF 13, 1Jul - 31 Dec 58, pp 10-11.

13. *Dept/State Bull*, 6 Sep 58, p 379; msg 2052, CNO to CINCSPECOMME, 24 Aug 58, in FRC; msg JCS 947046, JCS to CINCPAC & COMTDC, 25 Aug 58, in D/Plans, Ops 20-5, 22 Aug - 17 Sep 58.

14. Memo for COFSUSAF from Actg D/Plans, 28 Aug 58; memo for COFSUSAF from Asst/NSC affairs, subj: Probable Developments in the Taiwan Straits Area, 5 Sep 58, in D/Plans, Ops 20-9, 26 Aug - 18 Sep 58; msg 2465, Hq USAF to ATIC, ARDC, *et al.*, 11 Sep 58, in FRC:

15. Msg 2223, COMTDC to JCS, 31 Aug 58, in FRC.

16. Press Release, Text of Dulles statement on 4 September 1958 on Taiwan Situation.

17. Kuter Presentation; msg 2040, CINCPACAF to COFSUAF, 25 Aug 58, in FRC; Hist, I3th AF, 1 Jul - 31 Dec 58, pt III, p l; Special Rpt, Fifth Air Force in . . . Crisis, p 11; Hist, 12th AF, 1 Jul- 31 Dec 58, Pt III, p 20.

18. Interview, J. van Staaveren with Lt Col G. E. Tormoen, D/ops, 21 Jan 50.

19. Msg 3130C 58-206, 313th AD to 5th AF, App 2 in Special. Rpt, Fifth Air Force in . . . Crisis; Kuter Presentation; Chief of Staff's Policy Book, 1959, Item 140; Hist, 12th AF, 1 Jul - 31 Dec 58, Pt I, p 24.

20. CINCPACAF's Conclusions, in Hist, PACAF, I Jul - 31 Dec 58, Vol I, Pt I, pp 265-66; msg 2212, CINCPACAF to COFSUSAF, 31 Aug 58; msg 2211, CINCPACPAF to CINCPAC, 31 Aug 58; msgs 2224 & 2320, CINCPAC to JCS, 31 Aug & 6 Sep 58; msg 2330, Hq USAF to TAC *et al.*, 6 Sep 58, all in FRC; Hist, 12th AF, l Jul - 31 Dec 58, pt 1 pp 24 - 28.

21. Hist, 12th AF, 1 Jul - 31 Dec 58, Pt III, p 11.

22. *Ibid.* Pt I, p 28; Briefing by Col Wm. F. Barns, Lessons Learned - Taiwan Straits situation, at Senior Ops Officers Conf, 29-30 Jan 59, p 4; Chief of Staff's Policy Book, 1959, Items 140 & 140-2.

23. Hist, 12th AF, Pt I, p 28, & Pt III, pp 2, 20 - 21.

24. Hist, ADC, 1 Jul - 31 Dec 58, p 173, Barns Briefing, p 4, Kuter Presentation.

25. Hist, ADC, 1 Jul- 31 Dec 58, p 173, msg PACAF to 13th AF (personal Kuter to Moorman), 8 Sep 58, App 43 in Special Rpt, Fifth Air Force in . . . Crisis.

26. Hist , MATS, 1 Jul - 31 Dec 58, chap IV; Chief of Staff's Policy Book, 1959, Items 140 & 140-2.

27. See sources in n 26; msg 3295, CINCPAC to JCS, 15 Dec 58, in JCS files.

28. Anx B, in Hq PACAF, Report of Taiwan-Quemoy Operation, Dec 58, in D/Plans; Kuter Presentation; Air Force Operational Readiness Report, 29 Sep 58.

29. Msg 9876,COMSEVENTHFLT to CINCPACFLT, 4 Sep 58, in Marine Historical Archives (MHA); msg JCS 947046 as cited in n 13; Kuter presentation; Barns Briefing, p.5; *New York Times*, 17 Sep 58; *New York Herald Tribune*, 25 Sep 58; CINCPACAF's Conclusions as cited in n 20; msg 6359, CINCPACFLT to COMSEVENTHFLT, nd, in MHA; msg 6687, CINCPACFLT to COFSUSAF, 15 Sep 58, in D/Plans, Ops 20-5, 22 Aug – 17 Sep 58.

30. Msg JCS 947046 as cited in n 13; msg 7094, COMSEVENTHFLT to CINCPACFLT, 27 Aug 58; msgs, First MAW to CG FMFPAC, 011640Z & 170820Z, Sep 58; CO MAG-11 Rpt, 29 Aug - 11 Dec 58, p 21 Command Diary, Narrative Account of MAG-11, 1 - 30 Sep 58, Sec II, p 1, all in MHA; Anx B, pp 3 - 4, in PACAF Rpt of Taiwan - Quemoy Op.

31. MAG-11 Comd Diary, Sep 58, App G; msg, CG FMFPAC to CMC, 26 Aug 58; msg, First MAW to CG FMFPAC, 170820, Sep 58; CO MAG-11 Rpt; p 12, all in MHA.

32. MAG-11 Comd Diary, Sep 58, Sec 11, p 1; msg, CINCPACFLT to CG FMFPAC, 11 Sep 58, in MHA.

33. Msg 1297, COMSEVENTHFLT to COMPACFLT 081350Z Sep 58; MAG-11 Comd Diary, Sep 58, Sec II, p 5, & Oct 58, pp 1, 12 – 14; CO MAG-11 Rpt, pp 14 – 22; msg; First MAW to CG FMFPAC, 170820Z Sep 18, all in MHA.

34. Msg JCS 947046 as cited in n 13; Kuter Presentation; Hist, ATF 13, 1 Jul - 31 Dec 58, p 17.

CHAPTER III

1. Hist Div, 13th AF, Special Rpt, Taiwan Straits Operations, pp 30 - 31, 117, 158; msg 2552, CINCPACAF to COFSUSAF, 18 Sep 58, in FRC; Hq 313th AD G0 33, 13 Oct 58, App 9 in Special Rpt, Fifth Air Force in the Taiwan Straits Crisis of 1958.

2. Hq PACAF, Report of Taiwan-Quemoy Operation, Dec 58, p 6; interview, J. Van Staaveren with Lt Col G. E. Tormoen, D/Ops, 21 Jan 60; msg JCS 947046, JCS to CINCPAC & COMTDC, 25 Aug 58; 13th AF Rpt, Taiwan Straits Ops, p 224; msg 9937, CINCPAC to CNO, 5 Sep 58, in D/Plans, Ops 6-33, Sec 2.

3. Msg 2051, CINCPAC to CINCPACFLT *et al.*, 25 Aug 58, in FRC.

4. Msg 2076, CINCPAC to JCS, 26 Aug 58; msg 2064, CINCPACAF to COMUSJAPAN, 26 Aug 58; msg 2085, CINCPACAF to CINCPAC, 26 Aug 58, all in FRC.

5. Msg 2159 (JCS 247298), JCS to CINCPAC & COMTDC, 29 Aug 58, in D/Plans; memo for Gens Taylor, White, Pate, & Adm Burke from Chmn, JCS (Twining), subj: Authority for Emergency Action on Defense of Taiwan and the Offshore Islands, 15 Sep 58, in D/Plans, Ops 20-9, 26 Aug - 18 Sep 58, Sec 2.

6. Msg 0243, CINCPAC to JCS, 5 Sep 58, in FRC.

7. Anx H, 10 Sep 58, to CINCPAC Ops Plan 25-58; Anx N, 15 Oct 58, to CINCPACAF Ops Plan 25-58, both in D/Plans; msg 2450, CINCPAC to ALSUBCOM, 11 Sep 58, in FRC.

8. Msg 2450, CINCPAC to ALSUBCOM, 11 Sep 58; msg, CINCPACAF to COFSUSAF, 6 Sep 58, to D/Plans, Ops 6-33; msg 2437 (JCS 947808), JCS to COMNAVPHIL, Snagly Point RP, 10 Sep 58, in FRC.

9. Anx H, CINCPAC Ops Plan 25-58.

10. *Ibid.*

11. *Ibid.*; Taiwan file, in MHA; msg 2157, Hq USAF to ALMAJCOM, 29 Aug 58, in FRC; Hist, ATF 13, 1 Jul - 31 Dec 58, pp 19-20; Anx B to PACAF Rpt of Taiwan-Quemoy Op.

12. Hist, ATF 13, 1 Jul - 31 Dec 58, pp 19-20, 129-30.

13. Barns Briefing, p 3, & Tacon Briefing, pp 9-11, both at Senior Ops Officers Conf, 29 - 30 Jan 59; Hist, 12th AF, 1 Jul - 31 Dec 58, Pt 1, pp 37-38.

14. 13th AF Rpt, Taiwan Straits Ops, pp 63 – 64; Hist, 12th AF, 1 Jul – 31 Dec 58, Pt I, pp 37-38.

15. Anx F, 20 Aug 58, to CINCPACAF Ops Plan 25 - 58; Briefings for Gen Kuter by Lt Col Wm. D. Morrison, 26 Sep and 5 Nov 58, Apps 47 and 48 in Rpt, Fifth Air Force in . . . Crisis; Kuter presentation, 20 – 21 Nov 58; AU Evaluation Staff Study of Taiwan Crisis, dtd 1959; 13th AF Rpt, Taiwan Straits Ops, p 75.

16. 13th AF Rpt, Taiwan Straits Ops, p 75; Anx C, pp 10-11, PACAF Rpt of Taiwan-Quemoy Op; Briefings for Kuter by Morrison.

17. AU Eval Study of Taiwan.

18. PACAF Rpt of Taiwan-Quemoy Op, p 17; Hist, MATS, 1 Jul - 31 Dec 58, Chap IV.

19. Kuter Presentation; 13th AF Rpt, Taiwan Straits Ops, pp 79-80; AU Study of Taiwan; PACAF Rpt of Taiwan-Quemoy Op; pp 17 – 19 Briefing by Tac Div, D/Ops, USAF Lessons Learned in Lebanon and Taiwan-1958, p 40.

CHAPTER IV

1. Msg 2159 (JCS 247298), JCS to CINCPAC, 29 Aug 58; Anx H to CINCPAC Ops Plan 25-58, both in D/plans.

2. Baltimore Sun, 8 Sep 58; msgs 2484 & 2586, Hq USAF to ALMAJCOM, 12 & 19 Sep 58, both in FRC.

3. Msg, CINCPAC to JCS, 150127Z Sep 58, in D/Plans; msg 2295, Hq USAF to ALMAJCOM, 4 Sep 58, in FRC; PACAF Rpt of Taiwan-Quemoy Op, p 6; DAF Journal of Mutual Security, XII (Nov 58), p 67.

4. AIIS IR-789-58, subj: Fighter Aircraft Performance and Tactics in Air Battle of 24 September 1958; 3 Oct 58; AIIS IR-734-58, subj: Resume of CHICOM and CHINAT Participating Pilots, 7 Oct 58; Hist, ATF 13, 1 Jul - 31 Dec 58, p 26; msg CNO/MAAG to CNO, 1 Sep 58, in FRC; AIIS IR-73I-58, subj: Fighter Aircraft performance and Tactics in Air Battle of 10 Oct 58, 14 Oct 58.

5. AIIS IR-734-58 7 Oct 58; Fifth Air Force Commander's Briefing by Capt G. N. Earnhart, Jr, 25 Sep 58, Doc 15 in Special Rpt, Fifth Air Force in the Taiwan Straits crisis of 1958; AU Evaluation staff study of Taiwan Crisis.

6. Memo for DCS/P&P from Dep D/War Plans, subj: CNAF Reconnaissance Capability, 27 Aug 58, in D/Plans, Ops 20-5, Sec 4; *Washington Post and Times Herald*, l6 Oct 58.

7. Rpt, Fifth Air Force in . . . Crisis, pp 15, 81-82; Hist, ATF l3, P 26; memo for Gens Moore, Gerhart, Roberts, and Adm Dennison, Subj: Improvement of Chinese Nationalist Air Force Capability, 18 Aug 58, in D/Plans, Ops 20-5, 5 Jun - 21 Aug 58.

8. Memo for JCS from Dept/Navy, subj: Resupply of Govt of Republic of China-Held Offshore islands, 15 Sep 58, to D/Plans, Ops 20-5, Sec 4; msg, COMDC to CINCPAC, 21 Sep 58, in FRC; msg 2159 as cited in n 1; msg 8090, COMSEVENTHFLT to CTF 72, 30 Aug 58, in MHA.

9. Msg 2248, CINCPAC to COMTDC, 1 Sep 58; msg 2276, COMTDC to CNO, 3 Sep 58, both in FRC; *New York Times*, 4 Sep 58; *New York Herald Tribune*, 4 Sep 58.

10. Msg, COMTDC to CINCPAC, 6 Sep 58; msg 2518, COMTDC/MAAG to CNO 14 Sep 58, both in FRC; memo for JCS from Dept/Navy, 15 Sep 58, as cited in n 8; msg, CINCPACAF to COFSUSAF' 24 Sep 58, in D/Plans, Ops 20-5, Sec 5; msg 6830, COMTDC to CINCPAC, 25 Sep 58, in MHA; msg 400, Taipei to Secy/State, 15 Sep 58, & msg 2586, Hq USAF to ALMAJCOM, 19 Sep 58, both in FRC.

11. Msg 2276, COMTDC to CNO, 3 Sep 58; msg 2303, CINCPAC to JCS, 5 Sep 58; msg 2494, CNO to COMTDC & CINCPAC, 12 Sep 58, msg 2482, JCS to CINCPACAF *et al*., 12 Sep 58, all in FRC.

12. Memo of Conversation, subj: The Taiwan Strait Crisis, Dept/State (Dulles, Twining, *et al*.); 20 sep J8, in D/Plans, Ops 20-9; msg 2629 & 2630, CINCPAC to JCS, both 22 Sep 58; msg 2694, Hq USAF to ALMAJCOM, 27 Sep 58; msg 2725, COMTDC/MAAG to CINCPAC, 27 Sep 58, all in FRC.

13. Msgs 2586 & 2694, Hq USAF to ALMAJCOM, 19 & 26 Sep 58, msg 2629, CINCPAC to JCS, 22 Sep 58; msg 2705, CINCPAC to COMTDC, 26 Sep 58, all in FRC; Rpt, Taiwan Straits Crisis, by Col H. B. Hull, Doc 1 in Hist, PACAF, 1 Jul - 31 Dec 58, Vol II.

14. Hull Rpt, Taiwan Straits Crisis; msg 2790, COMTDC to CINCPAC, 4 Oct 58, in FRC; msg 6432, CINCPAC to COMTDC *et al.*, 24 Sep 58, in MHA.

15. Msg 2817, COMTDC to CINCPAC, 8 Oct 58, in FRC.

16. Msg, COMTDC/MAAG to CINCPAC, 2 Oct 58, in FRC; *New York Times* 6-7 Oct 58; *New York Herald Tribune*, 6 Oct 58; Hist, ATF 13, 1 Jul – 31 Dec 58, p 11.

17. Msgs 2853 & 2896, Hq USAF to ALMAJCOM, 10 & 17 Oct 58, in FRC; *New York Times*, 13 Oct 58.

18. Msg 0853, COMTDC to CINCPAC, 9 Oct 58, in FRC.

19. Memo for Asst SOD (ISA) from Chief, Mutual Security Div, DA, subj: Provision of 240mm Howitzers to Taiwan under MAP, 29 Oct 58, in D/Plans, Ops 20-5; memo for Twining (JCS) from Taylor, 22 Oct 58, in D/Plans, Taiwan 01 & 02.

20. *Washington Post* and *Times Herald* 24 Oct 58.

21. *New York Herald Tribune*, 20 Oct 58; memo for Twining (JCS) from Taylor, 21 Oct 58, in D/Plans, Taiwan 01. & 02; *New York Times*, 25 Oct 58; msg, COMTDC to CINCPAC et al., 3 Nov 58, in FRC.

22. *New York Times*, 25 Oct 58; msg, COMTDC to CINCPAC et al., 3 Nov 58; DAF Journal of Mutual Security, XII (Nov 58), p 67.

23. Msg 3000, CNO to CINCPAC, 30 Oct 58, in FRC.

24. Msg 3158, CINCPAC to CNO, 22 Nov 58; msg, Hq AFMTC to Hq USAF (personal Kuter to Gerhart), 23 Nov 58, both in FRC.

25. Msg, CINCPACAF to COFSUSAF, 16 Nov 58, in D/Plans, Ops 20-6, Sec 5; msg, Kuter to Gerhart as cited in n 24; msg 3162, Hq AFMTC to Hq USA. (also Kuter to Gerhart), 23 Nov 58, in FRC; memo for COFSUSAF from D/Plans, subj: Prepared JCS Reply to CINCPAC Msg 22 Nov 58, in D/Plans, Ops 20-5, Sec 6; msg 3158 as cited in n 24; msg 3245, Hq USAF to COMTAC, 5 Dec 58, in FRC.

26. Msg 3203, CINCPAC to COMTDC, 27 Nov 58; msg 3206, COMTDC/MAAG to CINCPAC, 28 Nov 58, both in FRC.

27. Hist, 12th AF, 1 Jul - 31 Dec 58, Pt III, pp 29-30.

28. *Ibid.*, Pt I, pp 42 43; Pt III, pp 11, 30-32.

29. *Ibid.*, Pt IIII pp 32 - 34 43.

30. Rpt Fifth Air Force in . . . Crisis, pp 94 – 95; AF Opnl Readiness Rpt, 29 Dec 58; Hist, ATF 13, 1 Jan -30 Jun 59, p 16; msg 3158 & msg from Kuter to Gerhart both as cited in n 24 memo prep by D/Plans, subj: Disposition and Redeployment of USAF Forces Committed to Support Taiwan, 25 Nov 58, in D/Plans, Ops 20-5, Sec 6; MAG-11 Command Diary, 1 Feb - 16 Mar 59, App D, in MHA.

31. Hist, Asst/Mutual Security, 1 Jul - 31 Dec 58, Sec 4, memo for D/Plans from Asst D/Policy, D/Plans, subj: Improvement of Chinese Nationalist Air Force capability, 15 Jan 59, in D/Plans, Ops 20-5, sec 6; memo for DCS/P&P from Actg D/Plans, 9 Apr 59, in D/Plans, RL (59) 244 Foreign Aids -- China, Sec 2; msg, CINCPACAF to COFSUSAF, 15 Nov 58, in D/Plans, 0ps 20-5, Sec 5, msg 3158 & msg from Kuter to Gerhart both as cited in n 24.

CHAPTER V

1. CINCPAC Ops Plan 25-58, 15 May 58.

2. Kuter Presentation, 20-21 Nov 58; AU Evaluation Staff Study of Taiwan Crisis; 13th AF Rpt, Taiwan Straits Ops, p 28.

3. File Ops 22, 1 Jan 59, in D/Plans.

4. Hist, PACAF, 1 Jul - 31 Dec 58, Vol I, Pt 1, p 167.

5. *Ibid.* pp 265-72, Hist, ATF 13, 1 Jul - 31 Dec 58, p 131; AU Eval Study of Taiwan; 13th AF Rpt, Taiwan Straits Ops, p 222, Kuter Presentation.

6. Interview, J. Van Staaveren with Lt Col G. E. Tormoen, D/Ops, 21 Jan 60.

7. 13th AF Rpt, Taiwan Straits Ops, pp 222-23.

8. Ltr, Kuter to LeMay, 31 Dec 58, in D/Plans, Ops 20-9, Sec 4; Kuter Presentation; AU Eval Study of Taiwan; Hist, 12th AF, 1 Jul - 31 Dec 58, Pt III, p 13; 13th AF Rpt, Taiwan Straits Ops, p 220; Tacon Briefing, at Senior Ops Officers Conf, 29 - 30 Jan 59, pp 7-8; msg, CINCPAC to JCS, 15 Dec 58, in JCS files.

9. Msg 0243, CINCPAC to JCS, 5 Sep 58, in FRC; msg, CINCPAC to JCS, 15 Dec, & 1tr, Kuter to LeMay, 31 Dec, both as cited in n 8; file Ops 22, 1 Jan 59.

10. Msg, CINCPAC to JCS, 15 Dec, & ltr, Kuter to LeMay, 31 Dec, both as cited in n 8.

11 Hist, ATF 13, 1 Jul - 31 Dec 58, p 127.

12. Msg, CINCPAC to JCS, 15 Dec 58; msg, COMTAC to COFSUSAF, 16 Dec 58, in D/Plans, Ops 20-9, Sec 4, Kuter Presentation; Hist, PACAF, 1 Jul - 31 Dec 58, Vol. I, Pt 1, p 270.

13. CO MAG-22 Rpt, 29 Aug - 11 Dec 58, p 22, in MHA.

14. File Ops 22, 1 Jan 59.

15. Kuter Presentation; AU Eval Study of Taiwan; file Ops 22, 1 Jan 59.

16. File Ops 22, I Jan 59; AU Eval Study of Taiwan.

17. 13th AF Rpt, Taiwan Straits Ops, p 70.

18. Memo for DCS/P&P from Dir/Info Svs, subj: Public Information Policies 17 Sep 59, in D/Plans, RL (59), Policy toward Taiwan.

19. Msg, CINCPAC to JCS, 15 Dec 58; AU Eval Study of Taiwan; file Ops 22, 1 Jan 59; 13th AF Rpt, Taiwan Straits Ops, p 228.

20. Msg, CINCPAC to JCS, 15 Dec 58, file Ops 22, 1 Jan 59, Hist, PACAF, 1 Jul - 31 Dec 58, Vol I, Pt 1, p 268.

21. Hearings Before the Subcmte of the House Cmte on Appropriations, 86th Cong, 1st Sess, *DOD Appropriations*, 1960, Pt 1, pp 894 - 95.

GLOSSARY

AACS	Airways & Air Communications Service
ACW	Aircraft Control and Warning
AFCIN	Assistant Chief of Staff, Intelligence, Hq USAF
AFMTC	Air Force Missile Test Center
AIIR	Air Intelligence Information Report
ALMAJCOM	All Major Commands
Anx	Annex
AOC	Air Operations Center
ATF 13 (P)	Air Task Force 13 (Provisional)
ATIC	Air Technical Intelligence Officer
BLT	Battalion Landing Team
CASF	Composite Air Strike Force
CCC	Combat Control Center
CINCAMBRITFOR	Commander-in-Chief, American-British Forces
CINCARPAC	Commander-in-Chief, Army, Pacific
CINCEUR	Commander-in-Chief, Europe
CINCFE	Commander-in-Chief, Far East
CINCLANT	Commander-in-Chief, Atlantic
CINCNELM	Commander-in-Chief, U.S, Naval Forces, Eastern Atlantic and Mediterranean
CINCPAC	Commander-in-Chief, Pacific
CINCPACAF	Commander-in-Chief, Pacific Air Forces
CINCSPECOMME	Commander-in-Chief, Special Command Middle East
CINCUSAFE	Commander-in-Chief, U.S. Air Forces in Europe
CINCUSAREUR	Commander-in-Chief, U.S. Army, Europe
CMC	Commander, Marine Corps
CNAF	Chinese Nationalist Air Force
CNO	Chief of Naval Operations
COC	Combined Operations Center
COMAMAIRFOR	Commander, American Air Forces
COMAMLANFOR	Commander, American Land Forces
COMAMNAVFOR	Commander, American Naval Forces
COMCIAPAC	Commander, Central Intelligence Agency Pacific
COMNAVPHIL	Commander, Navy, Philippines
COMSEVENTHFLEET	Commander, Seventh Fleet
COMTAIWANDEFCOM	Commander, Taiwan Defense Command
COMTASFME	Commander, Tactical Air Strike Force, Middle East
COMTDC	Commander, Taiwan Defense Command
COMUSJAPAN	Commander, United States in Japan
CSUSAF	Chief of Staff, USAF
CTF	Combined Task Force

DA	Department of the Army
DAF	Department of the Air Force
DEPTAR	Department of the Army
Dtd	Dated
EAME	European-African-Middle East
FEAF	Far East Air Forces
FEALF	Far East Air Logistics Force
FEC/UNC	Far East Command/United Nations Command
FMFPAC	Fleet Marine Force, Pacific
FRC	Federal Records Center
GCI	Ground Controlled Intercept
GRC	Government, Republic of China
HE	High explosive
JCC	Joint Communication Center
JOC	Joint Operations Center
MAAG	Military Assistance Advisory Group
MAG	Marine Air Group
MAW	Marine Air Wing
MATS	Military Air Transportation Service
MEAF	Middle East Air Force
MHA	Marine Historical Archives
MSTS	Military Sea Transportation Service
NAS	Naval Air Station
OCB	Operations Coordinating Board
OPLAN	Operations Plan
POL	Petroleum, oil, lubricants
RP	Republic of the Philippines
SITREP	Situation Report
TDC	Taiwan Defense Command
USAIRA	U.S. Air Attaché
USAREUR	U.S. Army, Europe
USARMA	U.S. Army Military Attaché
USEUCOM	U.S. European Command

APPENDIX ONE

SUMMARY 0F REFORT BY J-3 TO JCS ON LESSONS LEARNED FROM THE LEBANON and QUEMOY OPERATIONS WITH PARTICULAR REFERENCE TO LEBANON

The most detailed. analysis of the "lessons" of the Lebanon operation appeared in the JCS J-3 report approved 16 April 1959. These lessons were grouped under seven headings -- Politico-Military, Intelligence, Plans, Operations, Logistics, Communications, and Command -- and may be summarized as follows:

1. Politico-Military

a. The American public should be fully informed and conditioned as to the necessity of intervention in another country and the role of U.S. forces on such an occasion. More positive pronouncements are needed by political and. governmental leaders, During the Lebanon operation, a part of the press was apathetic.

b. Early action should be taken to establish a status of forces agreement when operations are to be conducted in a country whose sovereignty the United States recognizes. In Lebanon, the American Embassy was asked to negotiate such an agreement on 18 July, but despite the favorable attitude of the government it was not concluded until 5 August. Efforts at amendment were even slower.

c. Overflight and staging rights should be determined early for the benefit of planners and operators. In some cases it may be necessary to overfly without permission. Overflight problems with Austria, Switzerland, and Greece affected USAF operations.

d. If at all possible, the designated commander should be in the objective area prior to the beginning of operations in order to establish prompt contact with U.S. representatives and indigenous officials.

2. Intelligence

a. In joint operations, the services should establish common procedures for requesting and reporting reconnaissance and for use of agreed base maps. During the first operations in Lebanon aerial reconnaissance missions were conducted by the Navy under standard Navy and Marine Corps amphibious doctrine. With the arrival of the USAF tactical air units it became necessary to devise mutually agreeable procedures and draft a joint interim aerial reconnaissance plan.

b. Adequate maps and charts should be available in advance. These were initially inadequate.

c. Basic and current intelligence should be available at all command levels, Extensive deficiencies existed here.

d. Standardized procedures for reporting and disseminating intelligence should be established. Dissemination of intelligence on the political and military situation to subcommanders needed great improvement; on the other hand, reports by subcommanders were of substantial value.

e. Contingency plans should provide for necessary intelligence personnel and linguists to aid the commander, and joint service personnel should be specified by positions.

3. Plans

a. Formal check lists should have been maintained within the Joint Staff for actions required by Blue Bat. This would have improved Joint Staff operations during the earlier hours of the operation, when some confusion existed.

b. Interpretation of security restrictions by military commanders should be uniform, both to prevent disclosure of details of the operation and to lessen the possibility that other governments might miscalculate our intentions.

c. Planning must consider more closely the possibility of congestion at forward. airfields and provide for alternate air bases or phasing in of forces. Because the air base at Adana could not handle the peak loads, the arrival of the full USAF tactical air strength was delayed.

d. Planning for a special joint command should provide in detail for the organization and functioning of J-1. CINCSPECOMME Oplan 215-58 did not have a distinct administrative section.

e. Where ground, forces of more than one service are involved, an overall ground force headquarters should. be established. Representation should be specified by position, and personnel should be provided from the theater or the United States rather than from participating units. The joint ground force headquarters established on 26 July necessarily drew on participating units.

f. Plans for joint air operations should specify physical and procedural means for control and coordination. Joint armed forces doctrine giving guidance for all services participating in a given theater was not in existence.

g. Support forces should be closely fitted to combat forces and deployed as the situation dictates. Task Force Charlie, designed to support two battle groups, was airlifted to Lebanon in support of only one, thus reducing the available airlift for other units.

h. Airborne units should receive a periodic forecast of aircraft availability by type in order to keep loading tables and training exercises current. This should have facilitated deployment of the airborne task forces.

4. Operations

a. All unified and specified commands should be informed of impending operations in order to be ready to assume the proper alert posture for possible participation in the operation.

b. Amphibious operations should include sufficient tanks and helicopter support with the assault forces. The helicopters are needed to permit the rapid deployment of troops.

c. Adequate personnel and equipment for airport operations in the objective area should be provided.

d. A Joint Army - Air Force control group should be established at the principal departure and arrival airfields to coordinate loading, dispatch, unloading, dispersal, and services, In this way much of the confusion existing at Furstenfeldbruck and Adana could have been avoided.

5. Logistics

a. A more effective system of transition from "automatic" to "on call" resupply should be developed.

b. Plans for listing and outloading the initial sea-tail for the Army task force should be revised in order to avoid unnecessary delay in restoring operational readiness of ground force units after arrival in the objective area.

c. When procurement other than that associated with normal combat operations is required, personnel qualified in local procurement should be provided and planning should provide for contractual authority and funding by local commanders.

d. In medical planning for joint forces, responsibility for the provision of hospital support should be clearly fixed and the needs of all forces determined. USAREUR Emergency Plan 201 provided for hospitalization for Army units only.

e. The mounting-out plans for the Atlantic Fleet should provide for adequate Marine ammunition stocks. These were inadequate in the operation.

f. Planning should provide the proper offloading MSTS commercial ships where local labor is lacking or inefficient. Indigenous port labor in Lebanon was short because of the unsettled conditions.

g, The JCS commander must assume priority control over the movement of all aircraft assigned to him, with the JCS determining priority for the use of air bases in case of conflicts, as at Adana when tactical air units en route from the United States were forced to divert to Europe and North Africa. A Joint Military Transport Board would have provided better utilization of transport, particularly airlift.

h. It should be recognized that the timely employment of U.S. forces is dependent on strategically located base complexes and on adequate planning of logistic support. An additional base in Turkey may be necessary to support fixture planning.

i.   Lack of funds should not delay the issuance of necessary supplies in unprogrammed and unfunded operations.

6. Communications

a.   A detailed. low-classification theater-area communication SOP should be provided. The communication annex of the CINCSPECOMME plan had many provisions not known to lower echelons because of its Top Secret classification.

b. The capacity of the existing Army worldwide communication system for connecting with a deployed task force during a limited combat situation must be improved. During the Lebanon operation, long delays were not overcome until additional equipment was brought from the United States and. placed in operation.

c. To facilitate command control, additional equipment should be placed at strategic locations and maintained ready for use in case of tactical deployment.

d. Augmentation communication equipment should be propositioned at forward staging bases to reduce the need for airlift after the operation commences.

e. Adequate communication personnel and equipment should be provided for MSTS ships used for fleet marine forces.

f.   All aircraft participating in joint operations should be equipped to operate on common frequencies to facilitate traffic control and direction. Some aircraft were limited to VHF while overall operations were controlled by UHF.

7. Command

a.   Contingency plans must specifically delineate the authority and responsibility of the commanders concerned. Many questions of authority arose during the operation.

b.   Provision should be made for an orderly funding of joint emergency operations without major disruption of programmed activities.

Appendix Two
CASF X-Ray Tango Units And Their Home Stations

Parent Unit	Unit Deployed	Aircraft	Home Station
312th TF Wing	388th TF Sq	F-100	Cannon AFB NM
312th TF Wing	477th TF Sq	F-100	Cannon AFB NM
27th TF Wing	522nd TF Sq	F-101	Bergstrom AFB TX
345th TB Wing	499th TB Sq	B-57	Langley AFB VA
314th TC Wing	50th TC Sq	C-130	Sewart AFB TN
463rd TC Wing	773rd TC Sq	C-130	Ardmore AFB OK
837th AD	17th TR Sq	RF-101	Shaw AFB SC
837th AD	20th TR Sq	RF-101	Shaw AFB SC
4505 AR Wing	429th AR Sq	KB-50	Langley AFB VA
507th Tac Control Gp	Comm & Control ele		Shaw AFB SC
2d Tactical Depot Sq	Comd Ele		Langley AFB VA

APPENDIX THREE
USAF UNITS AVAILABLE TO PACAF AFTER DEPLOYMENT

Country	Bases	Unit	Aircraft	Mission
Taiwan	Taoyuan	83 FIS (NORAD)	12 F-104	Air defense
	Hsinchu	26 FIS (13th AF)	6 F-86D	Air defense
	Tainan	16 FIS (13th AF)	25 F-86D	Air defense
	Chiayi	388 TFS (CASF)	16 F-100D/F	Air defense & HE offense
	Tainan	868 TMS (13th AF)	20 TM-61	Offense (atomic)
	Tainan	507 Comm Ctrl		
	Kadena	Comm ele		
Philippines	Clark	1 WMT (CASF)		Atomic weapons maintenance
	Clark	26 FIS (13th AF)	7 F-86D	Air offense *
	Clark	72 TFS	25 F-100D/F	Offense
	Clark	50 TCS (CASF)	15 C-130	Airlift
	Clark	17 TRS (CASF)	6 RF-101	Recon
	Clark	4505 ARW (CASF)	5 KB-50	Refueling
Okinawa	Naha	25 FIS (5th AF)	29 F-86D	Air defense
	Kadena	522 TFS (CASF)	13 F-101	Offense
	Kadena	477 TFS (CASF)	16 F-101D/F **	Air defense
	Kadena	499 TBS (CASF)	14 B-57	Offense
	Kadena	20 TRS (CASF)	6 RF-101	Reconnaissance
	Naha	429 ARS (CASF)	5 KB-50	Refueling
Japan	Ashiya	773 TCS (CASF)	16 C-130	Airlift

* A typographical error for air defense in the original

** A typeographical error for F-100D/F in the original

www.ingramcontent.com/pod-product-compliance
Lightning Source LLC
La Vergne TN
LVHW091154080426
835509LV00006B/682